Rare Birds

Understanding and Addressing Air Force Underrepresentation in Senior Joint Positions in the Post–Goldwater-Nichols Era

Caitlin Lee, Bart E. Bennett, Lisa M. Harrington, Darrell D. Jones

Prepared for the United States Air Force
Approved for public release; distribution unlimited

For more information on this publication, visit www.rand.org/t/RR2089

Library of Congress Cataloging-in-Publication Data is available for this publication.
ISBN: 978-0-8330-9861-0

Published by the RAND Corporation, Santa Monica, Calif.

The RAND Corporation is a research organization that develops solutions to public policy challenges to help make communities throughout the world safer and more secure, healthier and more prosperous. RAND is nonprofit, nonpartisan, and committed to the public interest.

RAND's publications do not necessarily reflect the opinions of its research clients and sponsors.

Preface

General David Goldfein, chief of staff of the U.S. Air Force (USAF), has made the development of airmen for senior joint billets one of his top leadership priorities, announcing a series of broad efforts designed to increase Air Force representation in the most-senior joint positions. This report starts from the premise that the Air Force's first steps must be to identify which senior joint positions matter most and why, to understand how well the Air Force is represented in those positions, and to understand the extent to which the Air Force can actually control the factors that shape joint senior leader selection.

Our research findings indicate that the joint senior leader positions most widely viewed as critical are tied to strategymaking and warfighting. Among the most critical positions are chairman of the Joint Chiefs of Staff; vice chairman of the Joint Chiefs of Staff; combatant commands (particularly U.S. Central Command, U.S. Pacific Command, and U.S. European Command); joint task force commands; director of the Joint Staff; and the Joint Staff directors of the following three directorates: Operations (J3), Strategic Plans and Policy (J5), and Force Structure, Resources, and Assessment (J8). Further data analysis reveals that airmen are underrepresented in many of these key positions.

The USAF may be less competitive for senior joint positions for both subjective and objective reasons. Most notably, from a subjective perspective, interviewees perceived that the Air Force may have a cultural tendency to focus on grooming its rated force for top positions inside the service rather than systematically cultivating qualified officers for joint assignments. Objective considerations that may be reducing competitiveness include a potential shortfall in the quality of joint experience in terms of both Washington staff work and cross-domain exposure, a lack of joint experience early in airmen's careers, a lack of focus on strategic-level education focused on interagency cooperation and geographic expertise, and an inadequate organizational structure to support the establishment of joint task forces.

To adopt meaningful reforms that address these shortfalls, the USAF should consider whether it is willing to undertake a fundamental cultural transformation by taking on reforms that will effectively elevate the importance of senior joint command over senior Air Force command. Such reforms will require the Air Force to (1) openly examine and acknowledge its values and priorities in regard to senior leader development and (2) use those values and priorities as a basis for making conscious decisions about where to invest time and resources in joint senior leader development while acknowledging where corresponding trade-offs must be made.

This report is the result of the advice and assistance of a number of people. Although they are not named in the report, we owe special thanks to our interviewees for their candid discussion. We are also grateful to Col David Kumashiro and Katrina Jones of the Air Force General Officer Management Office (AF/DPG), who provided data and information on general officer management. Russell Frasz (AF/A1D) and Brig Gen Brian Killough (AF/A5S) and their staffs also were generous with their time and provided valuable insights on the Air Force's current and planned efforts to address the development of "jointness" within the Air Force. Finally, we would like to thank our reviewers, Paula Thornhill and Robert Elder, for their helpful and insightful feedback and recommendations.

This research is part of an ongoing RAND research effort to examine ways in which the Air Force could enhance the development and exposure of potential joint force commanders to make them more competitive for joint force senior leadership positions. The research reported here was commissioned by Air Force Manpower, Personnel and Services, Headquarters Air Force, and conducted within the Manpower, Personnel, and Training Program of RAND Project AIR FORCE as part of a fiscal year 2017 project, "Development of Joint and Combined Senior Leaders."

RAND Project AIR FORCE

RAND Project AIR FORCE (PAF), a division of the RAND Corporation, is the U.S. Air Force's federally funded research and development center for studies and analyses. PAF provides the Air Force with independent analyses of policy alternatives affecting the development, employment, combat readiness, and support of current and future air, space, and cyber forces. Research is conducted in four programs: Force Modernization and Employment; Manpower, Personnel, and Training; Resource Management; and Strategy and Doctrine. The research reported here was prepared under contract FA7014-16-D-1000.

Additional information about PAF is available on our website: www.rand.org/paf

This report documents work originally shared with the U.S. Air Force on May 1, 2017. The draft report, issued on May 11, 2017, was reviewed by formal peer reviewers and U.S. Air Force subject-matter experts.

Contents

Figures

Abbreviations

ACJCS	assistant to the chairman of the Joint Chiefs of Staff
CJCS	chairman of the Joint Chiefs of Staff
CJTF-OIR	Combined Joint Task Force—Operation Inherent Resolve
CSAF	chief of staff of the U.S. Air Force
DJ3	director of Operations
DJ5	director of Strategic Plans and Policy
DJ8	director of Force Structure, Resources, and Assessment
DJS	director of the Joint Staff
DoD	U.S. Department of Defense
FRE	Former Regime Elements
GNA	Goldwater-Nichols Act
GNO	Global Network Operations
JCS	Joint Chiefs of Staff
JIATF	Joint Interagency Task Force
JSOC	Joint Special Operations Command
JTF	joint task force
MDOCs	Multi-Domain Operations Centers
NAFs	numbered air forces
OSD	Office of the Secretary of Defense
PAF	RAND Project AIR FORCE
SAC	Strategic Air Command
SOJTF-A	Special Operation Joint Task Force—Afghanistan
USAF	U.S. Air Force
USAFRICOM	U.S. Africa Command
USCENTCOM	U.S. Central Command

USEUCOM	U.S. European Command
USNORTHCOM	U.S. Northern Command
USPACOM	U.S. Pacific Command
USSOCOM	U.S. Special Operations Command
USSOUTHCOM	U.S. Southern Command
USSTRATCOM	U.S. Strategic Command
USTRANSCOM	U.S. Transportation Command
VCJCS	vice chairman of the Joint Chiefs of Staff

1. Introduction

In his position as chief of staff of the U.S. Air Force (CSAF), Gen David L. Goldfein has identified the development of airmen for senior joint billets as one of his top priorities. In a September 20, 2016, speech, the chief argued that the Air Force, now more than ever, needs to be ready to stand up leaders and teams optimized for joint operations because airpower is poised to play a central role in all five of the main challenges confronting the U.S. Department of Defense (DoD) in China, Russia, Iran, North Korea, and violent extremist organizations.[1] To this end, General Goldfein has called for airmen to gain proficiency in joint warfare earlier in their careers so they can better contribute to campaign-level planning and for the Air Force to develop a core joint task force (JTF) staff that is ready to lead campaigns on short notice.[2]

General Goldfein is not alone in his desire to cultivate officers for the highest levels of joint command. Competition for the top joint positions is intense, owing to a military-wide recognition of a dramatic shift in power dynamics that began 30 years ago with the passage of the Goldwater-Nichols Act (GNA). Since the law was enacted in 1986, joint positions have subsumed a significant amount of authority historically reserved for the services. Under the law, the chairman of the Joint Chiefs of Staff (CJCS) became the principal military adviser to the President and the Secretary of Defense.[3] Previously, the service chiefs each had a direct line to these command authorities.[4] The law also empowered the combatant commands—which have been joint entities since 1946[5]—in two ways that solidified their supremacy over the formerly dominant military services.[6] First, the law gave the combatant commanders direct access to the Secretary of Defense but did not grant that privilege to the service chiefs.[7] Second, the law empowered the combatant commands to conduct current operations—a change that, in practice,

---

[1] Gen David S. Goldfein, remarks at the 2016 Air, Space and Cyber Conference, Gaylord Convention Center, Md., September 20, 2016a.

[2] Gen David S. Goldfein, *CSAF Focus Area: Strengthening Joint Leaders and Teams*, Washington, D.C.: U.S. Air Force, October 2016b.

[3] Joint Chiefs of Staff (JCS), *Joint Publication 1: Doctrine for the Armed Forces of the United States*, March 25, 2013, pp. III-4–III-5; Goldwater-Nichols Department of Defense Reorganization Act of 1968 (Pub. L. 99-433), Title II, Section 153.

[4] James R. Locher III, "Has It Worked? The Goldwater-Nichols Reorganization Act," *Naval War College Review*, Vol. 56, No. 4, Autumn 2001, p. 103.

[5] Cynthia Watson, *Combatant Commands: Origins, Structure, and Engagements*, Santa Barbara, Calif.: Praeger, 2011, p. 12–13.

[6] Paula Thornhill, *The Crisis Within: America's Military and the Struggle Between Overseas and Guardian Paradigms*, Santa Monica, Calif.: RAND Corporation, RR-1420-AF, 2016, pp. 3–5. Also see Locher, 2001, p. 108.

[7] Goldwater-Nichols Act of 1986, Title II, Section 162.

allowed them to become the fulcrums of warfighting, while the services were relegated to the role of force providers.[8]

As a result of these changes, each service views representation in the most-senior joint positions as one measure of its continued relevance, celebrating when it captures a key position and sounding alarms if it perceives an inequity.[9] In the U.S. Air Force (USAF), the professional literature reflects a particularly keen sensitivity to airmen's representation in senior joint billets. Airpower historians Philip Meilinger and Rebecca Grant, as well as retired Air Force Lt Gen David Deptula, have publicly expressed their concern that Air Force generals are "rare birds" in the most-senior joint positions and have each suggested that this constitutes a serious problem for U.S. national security that needs to be addressed.[10]

In 2009, senior Air Force leaders took up the issue, asking RAND to conduct an analysis of whether the USAF was actually underrepresented in the joint arena. That report found that between 2004 and 2008, Air Force officers were scarce in senior joint "warfighting" billets, including the combatant commands, as well as Joint Staff positions, including the J3 (Operations), J5 (Strategic Plans and Policy), and J8 (Force Structure, Resources, and Assessment) billets.[11] USAF representation in senior "non-warfighting" joint positions also declined during the same time period. Airmen competed poorly for these jobs, the study found, because they had insufficient joint warfighting experience by the time they were selected for brigadier general.[12] Another RAND study found that of the 90 JTF headquarters constituted

---

[8] Locher, 2001, p. 107; Watson, 2011, p. 18; also see Thornhill, 2016, p. 4, especially the Tommie Franks example.

[9] "Army Brass Losing Influence," Associated Press via Military.com, June 15, 2007; Harry Levins, "Pace Appointment Shows Marines Have Arrived," *St. Louis Post Dispatch*, April 23, 2005, p. 23; Michael Hoffman, "Obama Nominates First Airmen for JCS Leadership Position in 10 Years," *Military.com*, May 5, 2015; Scott Saslow, "Inside the US Navy's Leadership School," *Forbes*, April 27, 2010.

[10] Credit for the phrase "rare birds" goes to Rebecca Grant, "Why Airmen Don't Command," *Air Force Magazine*, March 2008, pp. 46–49. See also Col Phillip S. Meilinger, USAF (Ret.), "Airpower Past, Present and Future," presented at the Air Force Historical Foundation Symposium, Arlington, Va., October 16–17, 2007; and Gen David Deptula, USAF (Ret.), "Perspectives on Air Force Positions in Joint and COCOM Senior Leader Positions," presented at the Air Force Association Symposium, Orlando, Fla., February 22, 2013. Other airpower advocates who have written papers about the USAF's underrepresentation in senior joint billets: Maj William H. Burks (USAF), "Blue Moon Rising? Air Force Institutional Challenges to Producing Senior Joint Leaders," master's thesis, Fort Leavenworth, Kan.: School of Advanced Military Studies, U.S. Army Command and General Staff College, 2010; and Lt Col Howard D. Belote (USAF), *Once in a Blue Moon: Airmen in Theater Command—Lauris Norstad, Albrecht Kesselring, and Their Relevance to the Twenty-First Century Air Force*, Cadre Paper No. 7, Maxwell Air Force Base, Ala.: Air University Press, July 2000.

One oft-cited but misleading statistic in this literature, first cited by historian Phillip Meilinger, is that airmen served as theater commanders, heading JTF-like units, only four out of 110 times between World War II and 2008. But, given that the importance of joint capabilities is viewed as so much greater in the post-GNA environment than it was prior to the law, it may not be particularly helpful to consider USAF representation in the joint community prior to the GNA's enactment. See Meilinger, 2007.

[11] Al Robbert, "Air Force Competitiveness for Senior Joint Assignments: Recommendations for SECAF and CSAF," briefing, March 2009.

[12] Robbert, 2009.

between 1990 and 2008, the USAF formed the core of only 15 (17 percent), joining the other services to trail far behind the Army, which constituted the core of 47 JTFs (52 percent).[13]

As a practical matter, these concerns and findings support General Goldfein's interest in improving joint leader development. But viewed in a broader context, it is not immediately clear that the USAF's underrepresentation in the joint community, if it is still even statistically remarkable today, is a strategically significant problem—or one that the USAF alone can fix.

This report takes a top-down approach to help the USAF think through these broader strategic issues as it pursues its own bottom-up initiatives to reform joint leader development and JTF organization. Its purpose is to help the USAF consider which specific joint positions it should value the most and why; how well it is represented in those positions; and to what extent it can improve airmen's representation in those positions, given the wide array of variables that factor into the joint senior leader selection process. To be sure, there are some obvious things the USAF can do to better groom airmen for senior joint billets, such as General Goldfein's proposal to offer more opportunities to serve in joint positions earlier in airmen's careers. But whether even these changes are worth the time and effort requires a serious consideration of exactly what the USAF ultimately hopes to achieve by increasing its senior joint representation and contributions to JTFs. Does the USAF believe that the nation suffers when airmen are not at least equally represented in senior joint billets? If so, which billets most critically need the USAF's input? What would the USAF be willing to give up to assume greater representation in these positions? The USAF would likely benefit from considering each of these questions as part of a broader conversation about the rationale for making joint leader development a top priority *before* it exerts considerable time and resources toward reform of its joint senior leader development efforts.

Our findings suggest a consensus among senior military officers and DoD civilians (retired and active duty) that the USAF can best serve the nation if it targets a very specific set of senior joint jobs that play significant roles in strategymaking and warfighting. These current and former senior officials are more divided over whether it makes sense for the USAF to put a premium on JTF leadership, given the practical limitations of the USAF's existing organizational structure. Regardless, our data analysis suggests that the USAF has reason to worry about its representation in the joint positions if it agrees with interviewees' views regarding the positions that are most critical to the nation's warfighting apparatus. We found that the service is underrepresented in a range of positions that interviewees identified as key for strategymaking and warfighting (CJCS, director of the Joint Staff [DJS], director of Operations [DJ3], director of Strategic Plans and Policy [DJ5], and certain geographic combatant commander posts) and that the service has led only a handful of JTFs between 2005 and 2016.

---

[13] Michael Spirtas, Thomas-Durell Young, and S. Rebecca Zimmerman. *What It Takes: Air Force Command of Joint Operations*, Santa Monica, Calif.: RAND Corporation, MG-777-AF, 2009, p. 10.

That said, the USAF's prospects for improving its competitiveness for senior joint positions are limited by three considerations. First, the senior joint jobs that are most important for the nation change over time, to some extent, as the strategic environment evolves. Second, a wide variety of external factors beyond the USAF's control impinge on the decisionmaking process that surrounds senior joint leadership selection. Third, to the extent that the USAF can influence its competitiveness for the most-critical senior joint billets, there is a broader question that the USAF needs to consider about whether the Air Force is willing to take on some risk in the development of its officers for senior Air Force positions in order to support enhanced joint senior leader development.

Study Methodology, Limitations, and Framework

This report draws on the existing literature on joint leadership, current data on USAF representation in senior joint positions and JTFs, and project team interviews with active-duty and retired USAF general officers, retired U.S. Army and U.S. Navy general officers, and retired Office of the Secretary of Defense (OSD) civilians who were involved in the senior joint leader selection process. The discussion that follows contains references to the interviews, and, unless otherwise indicated, all quotations come from these interviews.

This report aims to spur a broader debate within the USAF regarding its goals for joint senior leader development and to offer broad recommendations for reforming the joint senior leader development process. It describes how various sources view the relative importance of senior joint positions and examines the USAF's representation in those positions. However, it does not make any judgments about which specific positions would most immediately benefit from increased Air Force representation, nor does it speculate on whether joint senior leader reforms would be worth the inevitable costs. The report also stops short of identifying specific actions the Air Force should take to increase its competitiveness for any particular senior joint position. Instead, our main goal is to help the Air Force think through what it hopes to achieve by reforming joint senior leader development and what trade-offs might be required. Once those decisions are made, further qualitative and quantitative analysis could be done to determine the career and education pathways that would increase the competitiveness of Air Force officers for the USAF's top priority joint senior positions.

The report proceeds as follows. Chapter 2 discusses how the GNA and/or changes in the international environment dramatically boosted the power and, therefore, the influence of three categories of joint positions: Joint Staff positions, combatant command positions, and JTF positions. Chapter 3 discusses how sources inside and outside the USAF rank the relative importance of senior joint positions and their rationale for these views, as well as these sources' views on whether the USAF should worry about competing for these positions. Chapter 3 also analyzes some basic data on USAF representation in the senior joint billets that matter most in the eyes of our interviewees. Chapter 4 turns to a discussion of the relative roles of the USAF and senior Pentagon leaders in the selection of joint senior leaders and discusses a variety of subjective and objective factors that shape both the USAF's competitiveness for senior leader selection and the senior leader selection process itself. Finally, Chapter 5 reviews findings and offers some broad recommendations for USAF leaders considering senior joint leader development reforms.

2. GNA and the Rise of the Joint Community

Joint positions have not always been important to the military services. In fact, the services railed against "jointness" in the four decades between the World War II–era establishment of the JCS and the unified commands and the 1986 passage of the GNA. Although Western militaries largely recognized unity of command as a virtue long before World War II, the services were hardly comfortable embracing it, for fear that centralized control would encroach on their basic roles and functions.[14] As a result of this "service-first" approach, the services and their backers in Congress effectively strived to keep joint institutions weak and to preserve a balance of power that favored the services prior to GNA reform.[15]

The advisory role of the JCS was minimal, as the law focused on a consensus approach that took into account all JCS views, leading to watered-down, late, and irrelevant advice to the President and Secretary of Defense.[16] The combatant commands held that status in name only. The four-star combatant commander, theoretically in charge of operations in his or her area of responsibility, did not have meaningful authority over the subordinate commanders, who, in practice, took orders from their own services rather than the combatant commander.[17]

As the administrative and operational shortcomings of the existing service-centric military organization became increasingly apparent, finally a service insider called for reform. USAF Gen David C. Jones, the CJCS, told a congressional panel in 1982 that service influence was so great, it would take outside pressure from Congress to achieve meaningful reform.[18] Over the next four years, lawmakers debated changes that ultimately led to the GNA. The law, signed by President Ronald Reagan on October 1, 1986, set the Pentagon on a course to reverse its service-dominant culture. It emphasized interservice cooperation at every level and across a broad swathe of military activity, a concept that became known as "jointness."[19] Unity of command was no

[14] The exception to this acceptance of unity of command might be the Navy, with its focus on independent command of ships at sea. See Carl Builder, *Masks of War*, Baltimore, Md.: Johns Hopkins University Press, 1989, p. 18. For the Western appreciation of unity of command dating back to the Napoleonic era, see Jay Kuvaas, ed., *Napoleon on the Art of War*, New York: Free Press, 1999, p. 63; and General Hermann von Kuhl (German Army), "Unity of Command Among the Central Powers," *Foreign Affairs*, September 1923. Regarding service worries about unified command, see Ronald H. Cole, Walter S. Poole, James F. Schnable, Robert J. Watson, and Willard J. Webb, *A History of the Unified Command Plan, 1946–2012*, Washington, D.C.: Joint History Office, Office of the Joint Chiefs of Staff, 2013, p. 1. See also Locher, 2001, pp. 95–98.

[15] Samuel Huntington, "Defense Organization and Military Strategy," *National Affairs,* No. 75, Spring 1984, p. 23.

[16] Locher, 2001, pp. 103–104.

[17] Locher, 2001, p.104, and Huntington, 1984, p. 24.

[18] Locher, 2002, pp. 33–58.

[19] Steven L. Rearden, *The Role and Influence of the Chairman: A Short History*, Joint History Office, Joint Chiefs of Staff, September 28, 2011, p. 454.

longer just an ideal; it was actually reflected in law and in the U.S. military's streamlined command structure. In contrast with the past, the services' instincts toward self-preservation were now held in check by a joint community empowered by GNA. In particular, the law both directly and indirectly enhanced the prestige and power of the CJCS, the Joint Staff, and the combatant commands.

The CJCS and Joint Staff, Empowered

The GNA designates the CJCS as the principal adviser to both the President and the Secretary of Defense. Under the GNA mandate, the CJCS can now provide a military-wide perspective.[20] The law also gives the CJCS the authority to provide strategic direction and conduct strategic planning and contingency planning; to provide advice on requirements, programs, and budget; to formulate doctrine, training, and education, and to make recommendations on roles and missions.[21] Although the CJCS is outside the chain of command (which runs directly from the President to the Secretary of Defense to the combatant commanders), the GNA empowers the CJCS to translate the commander-in-chief's intent into operational orders for combatant commands and to serve as a spokesman on behalf of the combatant commands to the national command authority.[22] In sum, the CJCS can now heavily influence both strategic and operational planning because of direct access to the President and Secretary of Defense, as well as the combatant commanders, in contrast with the limited access of the service chiefs.

The GNA also grants the CJCS a vice chairman of the Joint Chiefs of Staff (VCJCS), and authority over a Joint Staff of civilian and military personnel. Prior to GNA, the Joint Staff supported the work of all four service chiefs and the chairman.[23] The GNA also enhances the prestige of the Joint Staff relative to the service staffs, thanks to a provision requiring joint experience for the most-senior general and flag officer appointments.[24] In the years since the GNA was enacted, the influence of the Joint Staff has grown commensurate with its size. The Joint Staff has ballooned from 1,262 positions in fiscal year 2005 to a total of 2,599 positions in fiscal year 2013 (Joint Staff officials say that the increase is the result of the realignment of duties following the closure of Joint Forces Command).[25] Because of the size and competency of

---

[20] JP-1, p. III-4.

[21] JP-1, p. III-4–III-5, and Goldwater-Nichols Act, Title II, Sec. 153.

[22] JP-1, p. III-4, and Goldwater-Nichols Act, Title II, Sec. 163.

[23] Kathleen McInnis, *Goldwater-Nichols at 30: Defense Reform and Issues for Congress*, Congressional Research Service, R44474, June 2, 2016, p. 7.

[24] Rearden, 2011, p. 454.

[25] U.S. Government Accountability Office, *DoD Needs to Reassess Personnel Requirements for the Office of the Secretary of Defense, Joint Staff, and Military Service Secretariats*, GAO-15-10, Washington, D.C., January 2015, p. 9; McInnis, 2016, p. 7.

this highly professional staff, it has a unique capacity to frame choices in interagency discussions.[26]

Strategy, Foreign Policy, and Warfighting: The Role of the Combatant Commands Under the GNA

The GNA also dramatically enhances the influence, authority, and prestige of the unified combatant commands, to varying degrees. Today there are nine combatant commands. Six of them are considered "geographic commands" because they cover a specific area of responsibility in the world. They include U.S. European Command (USEUCOM), U.S. Pacific Command (USPACOM), U.S. Southern Command (USSOUTHCOM), U.S. Central Command (USCENTCOM), U.S. Northern Command (USNORTHCOM), and U.S. Africa Command (USAFRICOM). There are also three functional commands, including U.S. Transportation Command (USTRANSCOM), U.S. Strategic Command (USSTRATCOM), and U.S. Special Operations Command (USSOCOM), although the last command is better thought of as a hybrid, a concept discussed in more detail at the end of this chapter.

Combatant commands wield increasing influence on U.S. strategy and foreign policy both because of the GNA and because of changes in the international environment. Under the GNA, all combatant commanders now enjoy a direct line to the civilian leadership, allowing them to shape strategy more directly. But because of shifts in the strategic context since the end of the Cold War, geographic combatant commanders (as opposed to functional ones) have enjoyed a particularly broad expansion of their influence on strategymaking. The Cold War–era focus on a global warfighting scenario has given way to a regional focus tailored to the unique challenges presented by a multi-polar world. Geographic combatant commanders are therefore now in a unique position to use their regional expertise to shape strategy.[27] One example involves the near-closure of USCENTCOM in 1989. ADM William Crowe, the CJCS, prepared recommendations on national security strategy for Congress that omitted the Middle East. Worried about the fate of his command, U.S. Army GEN Norman Schwarzkopf came up with a new mission to save the command from becoming a backwater: checking Iraqi aggression against weaker states. With direct support from Secretary of Defense Dick Cheney, General Schwarzkopf effectively advocated to add the Middle East back into the national strategy.[28]

---

[26] Richard Kohn, "The Erosion of Civil Control in the United States Military Today," *Naval War College Review*, Vol. LV, No. 3, Summer 2002, p. 16.

[27] COL William W. Mendel, U.S. Army (Ret.), and Graham H. Turbiville Jr., *The CINC's Strategies: The Combatant Command Process*, Fort Leavenworth, Kan.: Foreign Military Studies Office, Combined Arms Command, December 1, 1997, p. 2.

[28] GEN Norman Schwarzkopf, U.S. Army (Ret.), *It Doesn't Take a Hero*, excerpt, *Newsweek*, September 27, 1992; "Oral History: Barnard Trainor," *PBS Frontline*, undated.

Because of their enhanced regional influence, combatant commanders can now shape not only strategic ends, but also ways and means—for better or worse. U.S. Army generals, for example, ran USCENTCOM from July 2003 to March 2007. While civilian leaders certainly shaped outcomes in Iraq, the Army's cultural preference for conventional combat also profoundly influenced the strategic environment.[29] As a result of their continued doctrinal focus on winning big conventional wars, Army leadership was slow to develop plans for Phase IV postcombat operations[30] and failed to release a counterinsurgency manual until 2008.[31] Similarly, while the Bush administration drew criticism for not immediately buying up-armored Humvees, the Army's own budget did not request those items, focusing instead on "big-ticket" purchases until the threat from improvised explosive devices could not be ignored.[32]

Geographic combatant commanders' enhanced regional influence also has allowed them to embrace a broader foreign policy portfolio. The responsibility for military engagement within a geographic combatant commander's theater has expanded to other traditionally civilian spheres, particularly as the State Department's budget has shrunk. The commander's "military diplomacy" role, which increased after the Cold War and accelerated after the September 11, 2001, attacks, includes the development of regional engagement strategies; efforts to build capacity in other countries; strategic information; and the disbursement of humanitarian, development, and security assistance.[33] Some authors have warned that geographic combatant commanders now have more influence than ambassadors and the State Department on foreign policy in their area of responsibility, particularly in the Pacific, Middle East, and Central Asia.[34] One recent study, involving interviews with two dozen ambassadors, found that these worries were exaggerated, but that there was a "troubling gap" between ambassadors, who are still the formal representatives in a region, and the combatant commanders, who increasingly wield more resources and a regional engagement agenda.[35]

In addition to expanding the geographic combatant commanders' roles in strategymaking and foreign policy, the GNA also enhanced all the combatant commanders' warfighting responsibilities. The GNA not only empowered combatant commanders to plan and oversee

---

[29] Mackubin Thomas Owens, "What Military Officers Need to Know About Civil-Military Relations," *Naval War College Review*, Vol. 65, No. 2, Spring 2012.

[30] Thomas E. Ricks, "Army Historian Cites Lack of Postwar Plan," *Washington Post*, December 25, 2004.

[31] Michael R. Gordon, "After Hard-Won Lessons, Army Doctrine Revised," *New York Times*, February 8, 2008.

[32] Owens, 2012, p. 76.

[33] Shoon Murray, "Ambassadors and the Geographic Commands," conference paper presented at Maritime Power and International Security EMC Chair Symposium, Newport, R.I.: US Naval War College, March 25–26, 2015. See also Derek S. Reveron, "Shaping and Military Diplomacy," presented at the 2007 Annual Meeting of the American Political Science Association, August 30–September 2, 2007.

[34] Dana Priest, *The Mission: Waging War and Keeping Peace with America's Military*, New York: W. W. Norton & Company, 2003, p. 71; Kohn, 2002, p. 17.

[35] Murray, 2015, p. 3.

military operations, but it also allowed these combatant commanders to establish whatever command and control arrangements they deemed appropriate for warfighting activities. In practice, this means that the combatant commands often stay focused on regional strategy and turn to subordinate units to deal with pressing issues for a finite period of time. One of the options at a commander's disposal is the creation of a JTF. The establishment of a JTF is a significant responsibility for the combatant commander. The decision to select an officer as a JTF commander is often a signal that he or she is destined for a fourth star, since the position offers valuable joint warfighting experience.

One clear pattern that emerges from this discussion is that two of the functional commands have experienced somewhat less growth in their influence over strategymaking, foreign policy, and warfighting. USSTRATCOM and USTRANSCOM are global, rather than regional, in nature, so they do not exert any particular influence on strategy or foreign policy in any one region of the world. And while functional commands can become the designated lead for military operations, they typically support the geographic combatant commander for regional military operations. To be sure, USSTRATCOM's expanded authority for space and cyber operations could lead to growing influence in the future, particularly if threats in those mediums continue to proliferate. Pending proposals to create full combatant commands for cyber and space would also potentially shift power dynamics. But at least for now, strategymaking and warfighting responsibilities at USSTRATCOM and USTRANSCOM are undertaken mostly in support of combat operations under the responsibility of the geographic commands.

The one exception to the functional commands' lower status is USSOCOM. In reality, USSCOM is a hybrid command, technically considered a functional command because of its global responsibilities but uniquely imbued with the authority to organize, train, and equip special operations forces across the service branches to serve under the geographic combatant commanders.[36] USSOCOM's strategymaking role is substantial. It has responsibility not only for special operations strategy, doctrine, and procedures, but it also serves as the lead combatant command for DoD planning against global terrorist networks.[37] In terms of foreign policy, the command has the authority to train troops in most countries, providing a quiet but powerful way to build relationships with other countries, even if diplomatic ties are tenuous.[38] Finally, in terms of warfighting, USSOCOM deploys forces in support of the geographic commands across the globe to conduct a spectrum of secret missions.[39] With legal responsibilities for military

[36] Andrew Feickert, *The Unified Command Plan and Combatant Commands: Background and Issues for Congress*, Washington, D.C.: Congressional Research Service, R42077, January 3, 2013, p. 15.

[37] Feickert, 2013, p. 15.

[38] Priest, 2003, p. 110–111.

[39] Priest, 2003, p. 32–33.

operations, intelligence activities, and covert action, USSOCOM's responsibilities also frequently overlap with those of the Central Intelligence Agency.[40]

[40] Thornhill, 2016, p. 4.

3. High-Value Targets: USAF Representation in the Senior Joint Positions That Matter Most

In the post-GNA world, the Joint Staff and the combatant commanders play critical roles. Bureaucratic politics tell us that military services, seeking to maximize their power and prestige, will seek to dominate important positions, or at least ensure equal representation. But not all joint positions are viewed equally. Through their own cultural lenses, each service evaluates these positions in terms of relative importance. Considerations may include the prestige, resources, or power associated with the position, as well as whether the service perceives that it can add value in the position, thereby maximizing its contribution to the joint fight.[41] A service's own views of the relative importance of a position gain further credibility if external actors—such as the other services and the senior civilian and military leaders responsible for joint selection—agree that the service's representation is critical to maximize U.S. warfighting capacity. This section explores whether there is any consensus between USAF and non-USAF sources about which senior joint positions matter most and whether the USAF's representation in these positions should be a primary USAF concern. It draws on interviews conducted with current and former senior civilian and uniformed officials who have been involved in the joint senior leader selection process.

Which Joint Positions Matter the Most?

Among interviewees inside and outside the USAF and in the existing literature, a consensus emerged that senior joint positions that involve strategymaking and warfighting are the most critical. A commonly shared view among the interviewees is that these positions are most important because they give the services the greatest opportunity to contribute to the defense of the nation. None of the interviewees mentioned material considerations, such as prestige, power, and resources, as primary factors in assessing the relative importance of senior joint positions, although they noted that these factors are often also present in the positions associated with strategymaking and warfighting.

Because of their influential leadership roles in strategymaking and warfighting, Joint Staff jobs, most specifically the CJCS, VCJCS, DJS, DJ3, and DJ5, and, to a lesser extent, the director of Force Structure, Resources, and Assessment [DJ8], are perennially important, according to interviewees. They further noted that the DJS, DJ3, and DJ5 are critical Joint Staff jobs not just in their own rights, but also because they are perceived as "upwardly mobile"—in other words,

[41] The idea that, in bureaucratic politics, organizations compete for roles and missions closest to their heart, or "organizational essence," comes from Morton Halperin, Priscilla Clapp, and Arnold Kanter, *Bureaucratic Politics and Foreign Policy*, 2nd ed., Washington, D.C.: Brookings Institution Press, p. 27.

"grooming grounds" for the most influential positions of CJCS, VCJCS, and combatant commander.

Not surprisingly, given their growing authority discussed in the last chapter, the geographic commands were also cited by many sources as being particularly important senior joint posts. In today's strategic context, many interviewees noted that the most important geographic commands are USCENTCOM, USEUCOM, and USPACOM because, as one retired Navy admiral involved in senior joint leader selection summed it up: "This is where the big wars happen; it's where the political stakes are higher, and you have to have somebody with more experience under their belt." However, several interviewees offered a major caveat: The importance of a given combatant command, geographic or functional, can change over time as the strategic environment shifts. For example, another retired Navy admiral involved in senior joint leader selection noted that USEUCOM was formerly a backwater, but that has changed with Russia's recent resurgence in the region.

Sources also largely agreed that, relative to functional commands, the geographic commands are generally more important, but this conclusion is context-dependent. For example, a former Secretary of Defense and a retired U.S. Army general who were both involved in joint leader selection noted that, during wartime in the Middle East, command of USCENTCOM has been viewed as critical, but it is difficult to say that the functional commands are therefore less important—particularly USSTRATCOM, given its central role in strategic deterrence, which is once again a salient issue in the changing 21st-century security environment.

Active-duty and retired USAF officers were less nuanced in their view of the importance of geographic commands relative to functional ones. One retired four-star USAF officer with significant functional command and joint experience argued that geographic commands are definitely more important, based on his own experience and his observations of congressional interest in the geographic commands, as opposed to the functional ones. Another active-duty general officer involved in senior joint leader selection shared that view. The relative lack of discussion of functional commands in the professional and academic literature seems to reflect the same implicit judgment that geographic commands are more important, at least in today's strategic context.[42]

Active-duty and retired USAF officers also agreed that JTF command and staff positions are important because of their lead operational roles. They argued that this has been the case since the enactment of GNA because JTFs are well understood to be important grooming positions for geographic combatant command. Two retired USAF generals noted that the most important JTFs for the USAF are those that are more air-centric than ground-centric (such as humanitarian relief

---

[42] Watson, 2011, does not discuss the functional commands. Other examples of literature that emphasize the power of the geographic commands and rarely, if at all, mention the functional commands include Edward Marks, "Rethinking the Geographic Combatant Commands," *InterAgency Journal*, Vol. 1, No. 1, Fort Leavenworth, Kan.: Colonel Arthur D. Simons Center for the Study of Interagency Cooperation, Fall 2010; and Priest, 2003. See also Grant, 2008; Meilinger, 2007; and Deptula, 2013.

missions that require delivery of goods over long distances) or combat operations with a significant air component (such as Combined Joint Task Force—Operation Inherent Resolve [CJTF-OIR]). The USAF must show an ability to lead those JTFs because the employment of strategic airpower is the USAF's chief responsibility. But one of those retired USAF generals who had experience in both the joint and functional command communities noted that it is also important for the USAF to be able to lead in complex mission environments that are not just exclusively associated with the aviation element, such as ground-based combat environments.

Should the USAF Worry About Its Representation in the Most Important Senior Joint Positions?

Sources inside and outside the USAF widely agreed that the USAF should make senior joint leader selection for critical positions a key priority because (1) the nation's warfighting capacity suffers if insufficient consideration is given to the air, space, and cyber realms at the most-senior levels of joint command, including the CJCS, VCJCS, and combatant commander positions, and (2) airmen are not likely to make it to the most-senior levels of joint command unless they serve in critical joint grooming positions within the Joint Staff or the combatant commands that are also important in their own right, such as the DJS, DJ3, or DJ5.

Geographic combatant commands are perceived as particularly important posts. "I'm always someone who believes you need a variety of views, and those three [USCENTCOM, USPACOM, and USEUCOM] are widely seen, and I think rightfully so, as the crown jewels in the military," said one retired Navy admiral involved in senior joint leader selection. "If you are going to be seen as a crown jewel, then you want to be represented, and that hasn't changed." Another retired U.S. Army general involved in senior joint leader selection echoed that view, noting that the USAF cannot dominate geographic commands in every cycle, but if the service is notably absent from certain billets for a prolonged period of time, it needs to investigate why that is the case.

Sources also cautioned, however, that the selection process for the commander of a geographic command is highly political, particularly in the case of USPACOM, traditionally dominated by the Navy with support from Congress. Therefore, while the USAF should put forth its best candidates, it should do so with the expectation that politics—and, specifically, biases for or against particular services, discussed further below—are highly influential. "The USAF can't worry about its geographic command reputation too much because there's so much politics that go into these decisions," said one active-duty general officer involved in senior joint leader selection. "The USAF should try to compete for USCENTCOM, but they can't expect anything because the perception is that USCENTCOM is all about the ground fights."

Airpower advocates and retired USAF officers echoed the concern that the airpower perspective is needed at the geographic commands,[43] which are seen as nerve centers for warfighting. "It's been hard to break the glass ceiling" at geographic commands like USCENTCOM, said one retired four-star USAF general with significant joint and functional command experience. "But it's important because among that group of shooters, the air and cyberspace perspective needs to have sufficient advocacy when you're devising military strategy for the country." A retired three-star USAF general with significant joint experience echoed that sentiment, arguing: "The geographic combatant commander sets the strategic narrative. If they are all from the ground forces, all the options are going to be ground-based. This is bad for the joint force; it's bad for the nation not to have a spectrum of perspectives at your combatant command."

Three retired USAF generals with significant joint experience mentioned USPACOM as a geographic command that would most benefit from USAF leadership. No airman has ever led the command, yet the challenges in that theater lend themselves to airpower. Two retired generals noted that USPACOM may be largely water, but it is "100-percent air." The retired USAF three-star general also mentioned USCENTCOM as another command that has never been led by an airman but could benefit from an airpower perspective, especially after 15 years of ground-focused conflict in that region led almost completely by ground-focused Army and Marine Corps generals.

Interviewees' views were more mixed regarding USSOCOM, the hybrid command. When asked to discuss the relative importance of the various combatant commands, interviewees almost never mentioned USSOCOM. This is somewhat surprising, given the amount of strategymaking and warfighting responsibility of the command. The lone exception to the silence was a retired Air Force four-star general who said that USAF representation in the senior leadership of USSOCOM was a high priority, particularly within the subunified command known as Joint Special Operations Command (JSOC), which controls the special mission units of USSOCOM, such as the Army's Delta Force, the Navy's Sea, Air, and Land (SEAL) Team Six, and the USAF's 24th Special Tactics Squadron.[44] One reason for reticence on the topic, at least among USAF interviewees, could be that special operations is largely conceived of as a functional specialty within the Air Force. Special operators pursue stovepiped career tracks that look very different from the career tracks of officers currently being groomed for senior positions in the USAF or joint community.

Interviewees inside and outside the USAF said that the USAF should also be concerned about its representation in JTF commander and staff positions because, like the geographic

---

[43] Grant, 2008, p. 49; Meilinger, 2007, slide 32; Belote, 2000, p. 2; Col Russell Mack (USAF), *Creating Joint Leaders Today for a Successful Air Force Tomorrow*, Maxwell Air Force Base, Ala.: Air University, May 11, 2010, p. 2.

[44] For a complete list of JSOC units, see John Pike, "Joint Special Operations Command (JSOC)," GlobalSecurity.org, undated.

commands, many of the most high-profile JTFs are viewed as warfighting organizations. "We should care about others' perception of the quality of Air Force combat leadership and our ability to lead in complex mission environments," said one retired USAF general with experience in the joint community and functional commands. JTF command positions are also important because the JTF commander can dictate strategy and the form and function of an operation and because JTF commander posts are viewed as grooming positions for future geographic combatant commanders, according to a retired USAF three-star general.

Geographic combatant commanders and JTF commanders are typically supported by an air component commander. But among interviewees, views are mixed about whether that position is as critical to strategymaking and warfighting as the previously discussed joint positions. The retired three-star USAF general argued that the air component is not the place to shape operations; air component commanders are limited to directing the air tasking order cycle, while it is the geographic combatant commanders and JTF commanders who are making the strategic decisions. But others in the USAF seemed to take the view that the air component commander is a highly significant position. Interviewees inside and outside the USAF said that the service currently has a tendency to groom its best airmen to be air component commanders rather than to lead JTFs or geographic combatant commands. This will be discussed more in the section below on how USAF culture shapes competitiveness on joint assignments.

What the Numbers Say

Having established the USAF's priorities when it comes to senior joint positions, we can now turn to a discussion of the data. What do the numbers show about USAF representation in the senior joint positions that some would say matter most? This section reviews airmen's representation in CJCS, critical joint staff positions, combatant commands, and JTF commands.

As Figure 3.1 indicates, since 1986, airmen have only held the office of the CJCS once, when USAF Gen Richard Myers manned the post from 2001 to 2005. His tenure accounted for 13 percent of this 30-year era, while Army generals occupied the position 51 percent of the time, Navy generals held the post 25 percent of the time, and U.S. Marine Corps generals held the position 11 percent of the time (Figure 3.1).

Figure 3.1. Secretary of Defense and CJCS Service Origins, 1986–2016

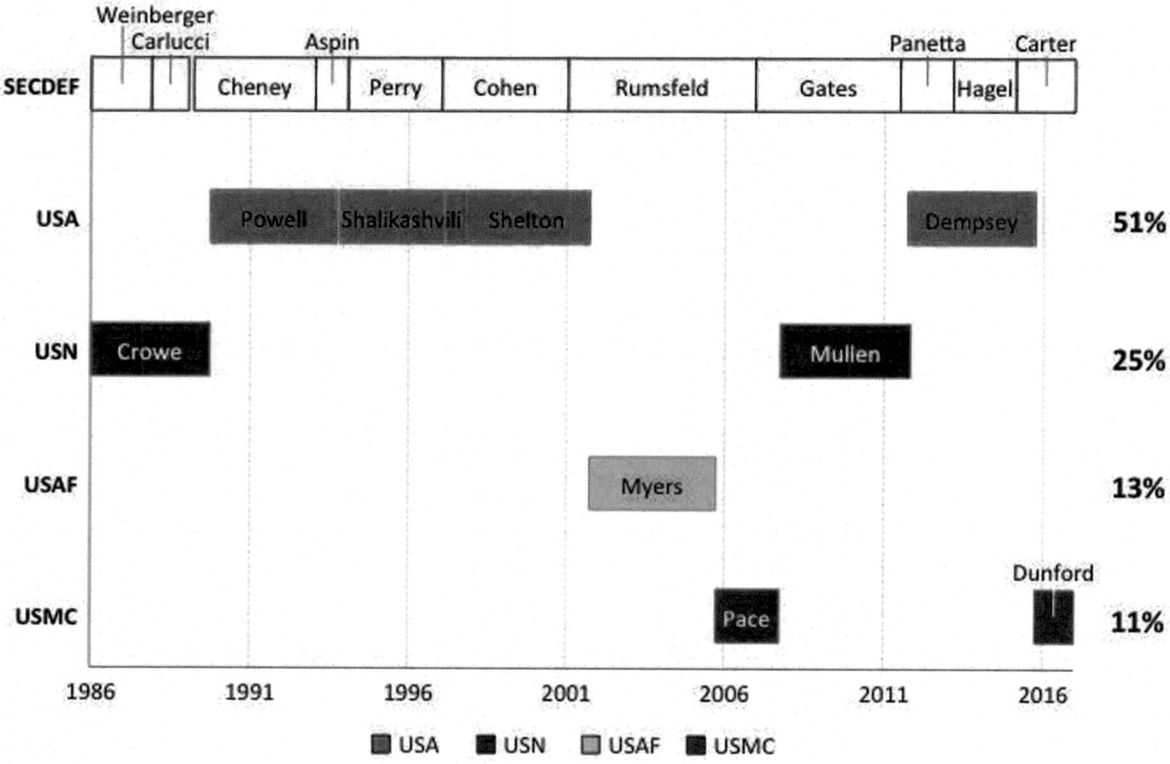

SOURCE: RAND analysis.

NOTES: SECDEF = Secretary of Defense; USA = U.S. Army; USN = U.S. Navy; USMC = U.S. Marine Corps.

The USAF has fared better in some Joint Staff positions, which are viewed as key joint grooming billets and influential positions in their own right (see Figure 3.2). Since 2005, the earliest year for which RAND could obtain complete data, airmen occupied the position of assistant to the chairman of the Joint Chiefs of Staff (ACJCS) 45 percent of the time and the J8 position 46 percent of the time. However, in the positions viewed as "four-star makers" and highly influential—the DJS, the DJ3, and the DJ5—the USAF's representation is far less robust. Airmen served as DJS 23 percent of the time, as DJ3 12 percent of the time, and as DJ5 only 8 percent of the time.

Figure 3.2. Critical Joint Staff Position Service Origins, 2005–2016

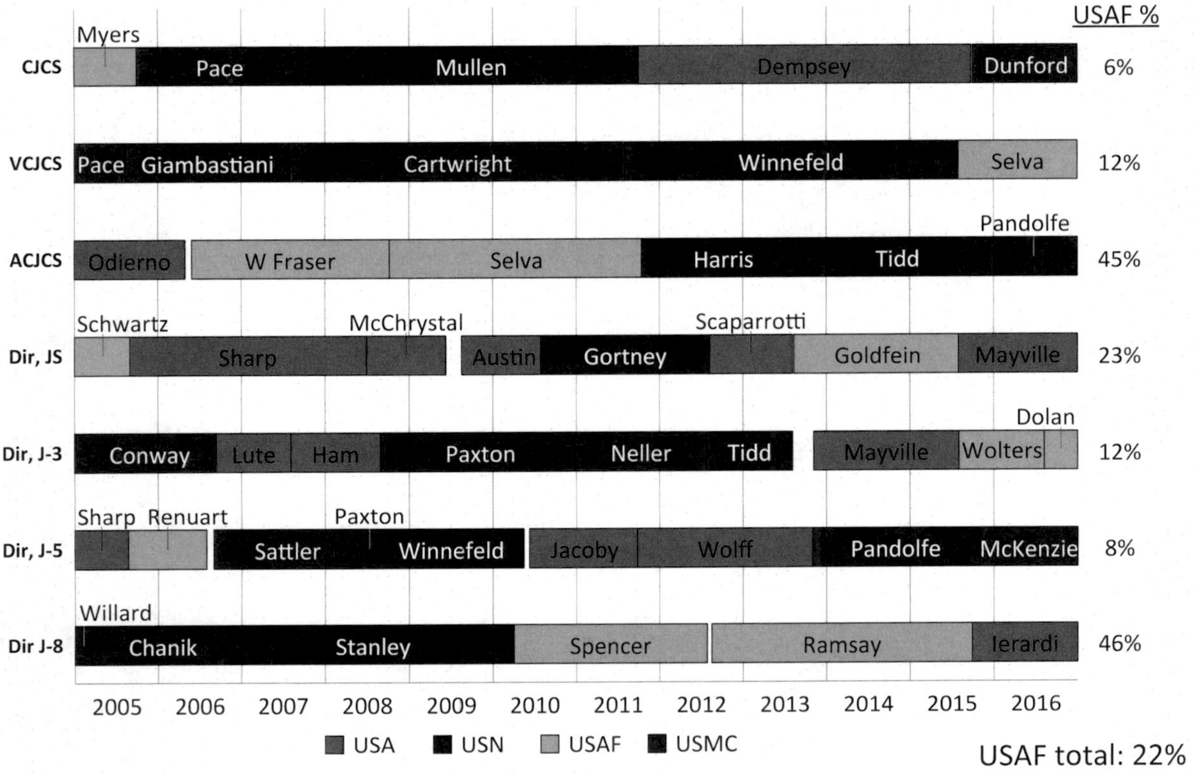

USAF %
CJCS: Myers, Pace, Mullen, Dempsey, Dunford — 6%
VCJCS: Pace, Giambastiani, Cartwright, Winnefeld, Selva — 12%
ACJCS: Odierno, W Fraser, Selva, Harris, Tidd, Pandolfe — 45%
Dir, JS: Schwartz, Sharp, McChrystal, Austin, Gortney, Scaparrotti, Goldfein, Mayville — 23%
Dir, J-3: Conway, Lute, Ham, Paxton, Neller, Tidd, Mayville, Wolters, Dolan — 12%
Dir, J-5: Sharp, Renuart, Sattler, Paxton, Winnefeld, Jacoby, Wolff, Pandolfe, McKenzie — 8%
Dir J-8: Willard, Chanik, Stanley, Spencer, Ramsay, Ierardi — 46%
2005 2006 2007 2008 2009 2010 2011 2012 2013 2014 2015 2016
USA USN USAF USMC
USAF total: 22%

SOURCE: RAND analysis.

The USAF also has struggled to occupy the combatant commander positions viewed as most critical since 1986 (Figure 3.3). On one hand, in terms of combatant commander positions overall, including the functional commands, the USAF is relatively well represented among the four service branches, occupying the combatant commander position 24 percent of the time (this assumes that the USMC gets an equal share of the pie, despite being a smaller service). However, if we focus on joint positions viewed as most critical, the USAF fares relatively poorly. Overall, the USAF leads the geographic combatant commands about 10 percent of the time, less than its fair share. Among those geographic combatant commands, the USAF has never commanded two that airpower advocates cite as being particularly critical: USCENTCOM and USPACOM. The service has never commanded USAFRICOM either. It does reasonably well at USEUCOM, occupying the commander's post 18 percent of the time, and slightly less well at USSOUTHCOM, which the service has commanded 12 percent of the time. Interestingly, the USAF has commanded USNORTHCOM 41 percent of the time, which, as mentioned above, is not surprising, given the command's focus on aerospace.

Figure 3.3. Combatant Command Leadership Service Origins, 1986–2016

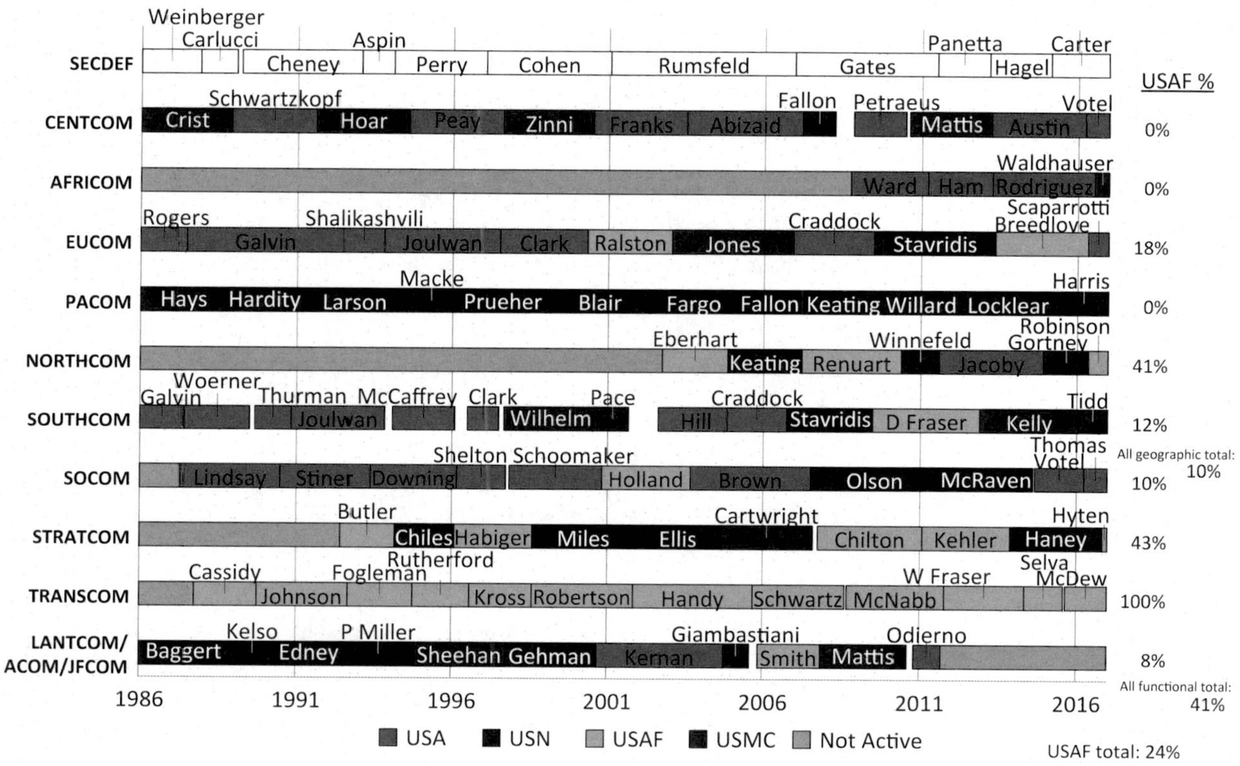

SOURCE: RAND analysis.

NOTES: ACOM = Atlantic Command; JFCOM = Joint Forces Command; LANTCOM = Atlantic Command.

In terms of JTF commanders, Figure 3.4 shows that the Air Force has not commanded the variety of high-profile warfighting JTFs established to confront terrorism and insurgency in the Middle East since the September 11, 2001, attacks. Figure 3.5 shows that the USAF has rarely commanded other types of JTFs that have been established since 2005. Our data set for Figure 3.5 includes JTFs that were constructed for a variety of purposes but does not include medical, Judge Advocate General, U.S. Coast Guard, or O-6-led JTFs. The Army and the Navy have staffed the lion's share of the JTF positions. The Air Force's longest opportunity to command JTFs is associated with the Global Network Operations (GNO) JTF that was subordinate to USSTRATCOM. USAF Brig Gen Michael Longoria served a six-month tour as the Joint Interagency Task Force (JIATF) Former Regime Elements (FRE). The USAF's most recent JTF commander, Maj Gen Scott Howell, recently took command of the Special Operation Joint Task Force—Afghanistan (SOJTF-A). In total, the Air Force commanded the JTFs only about 5 percent of the time.

In closing, a review of CJCS, critical Joint Staff positions, combatant command posts, and JTF command positions reveals that airmen are indeed underrepresented compared with the other services. The perennial concern of airpower advocates that airmen are "rare birds" in the senior joint positions that airpower advocates (and those outside the Air Force) viewed as most critical appears to be valid.

Figure 3.4. Service Origins of Commanders of High-Profile Warfighting JTFs Established Since 2005

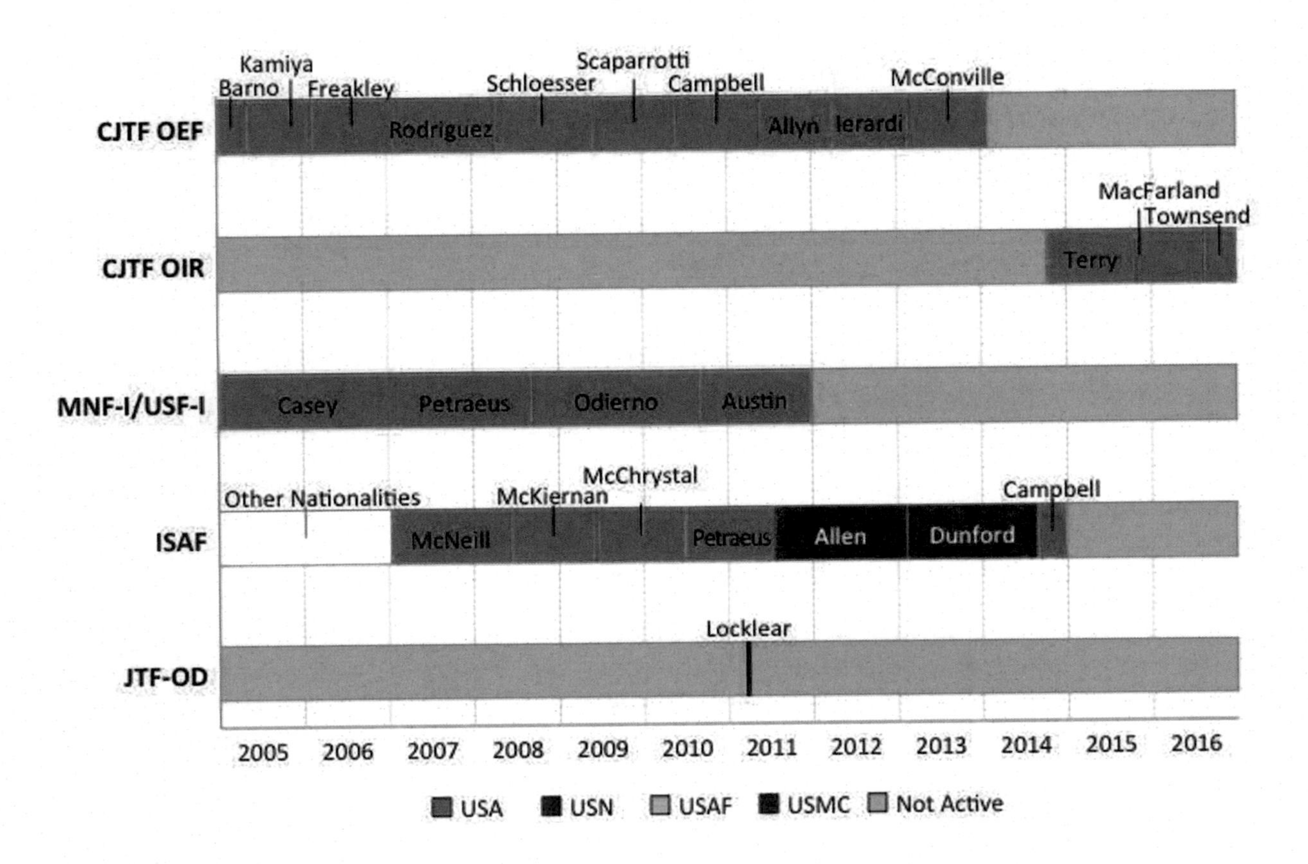

SOURCE: RAND analysis.

NOTES: The JTFs identified in this chart include Combined Joint Task Force Operation Enduring Freedom (CJTF OEF, Afghanistan); Combined Joint Task Force Operation Inherent Resolve (CJTF OIR, anti-ISIS campaign); Multi-National Force—Iraq/U.S. Forces—Iraq (MNF-I, USF-I, Iraq); International Security Assistance Force (ISAF, Afghanistan); and Joint Task Force Odyssey Dawn (JTF-OD, Libya).

Figure 3.5. Service Origins of Commanders of Additional Selected JTFs Established Since 2005

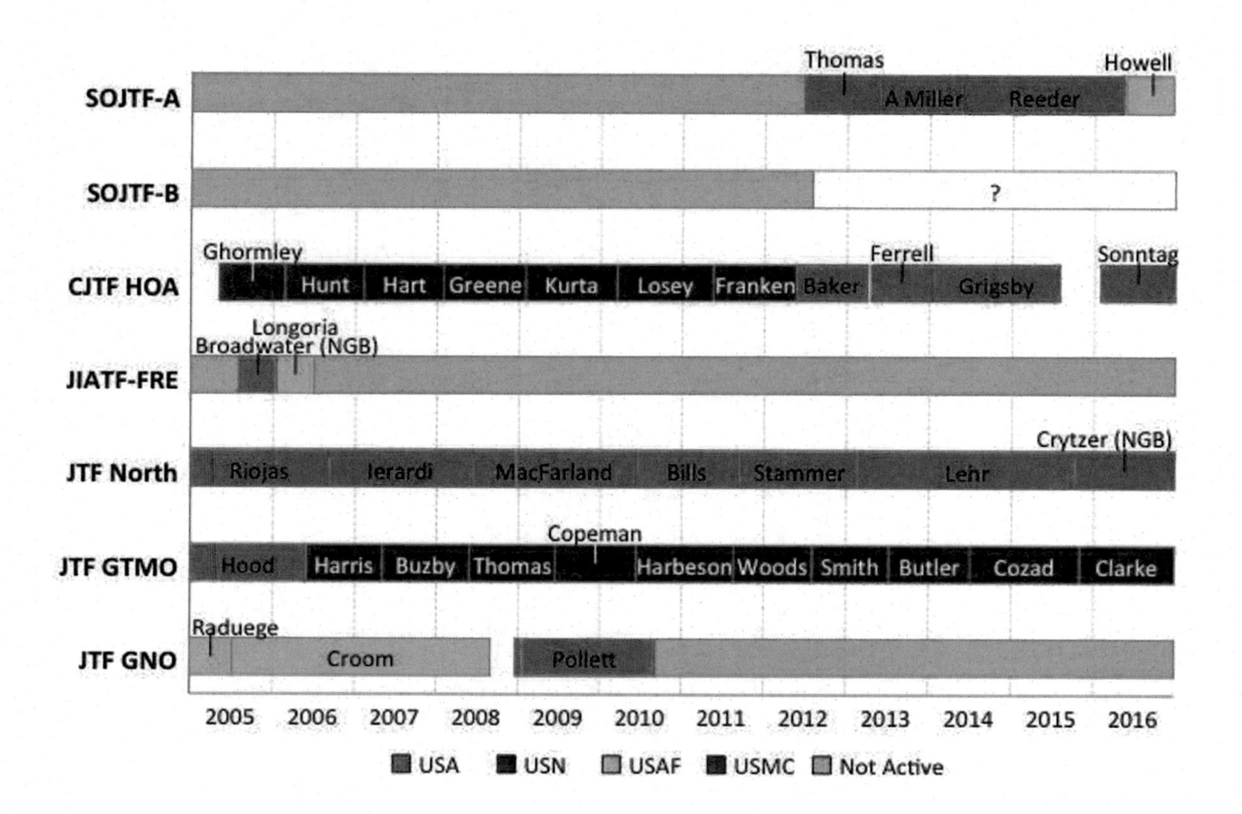

SOURCE: RAND analysis.

NOTES: The JTFs in this chart include Special Operations Joint Task Force—Afghanistan (SOJTF-A, Afghanistan); Special Operations Joint Task Force—Bragg (SOJTF-B, Fort Bragg); Combined Joint Task Force Horn of Africa (CJTF, HOA, Africa); JIATF-FRE, Iraq; JTF-North, United States; JTF Guantanamo Bay, Cuba; and JTF-GNO, United States.

4. Critical Factors That Influence Joint Senior Leader Development and Selection

If airmen are falling short in terms of representation in the CJCS position, Joint Staff positions, and geographic combatant command positions, why is this the case? One possibility is that the USAF is not offering qualified candidates for senior joint positions. Another possibility is that the decisionmaking process for senior joint leadership selection is shaped by a variety of factors that make it less likely for USAF officers to be selected for senior joint assignments. After interviewing a variety of senior officials who have been involved in the joint senior leader selection process both inside and outside the USAF, we found that aspects of both the Air Force's approach to senior leader development and the senior leader selection process itself have significantly affected the USAF's representation in the upper echelons of joint command.

This chapter will address each of these aspects in turn. First, it will address the subjective and objective factors that influence the Air Force's approach to joint senior leader development. Second, it will turn to a discussion of the joint senior leader selection process, including the subjective and objective factors that influence how that process works. By looking at both sides of the process—both development and selection—the goal is to give the USAF some insight into how it might adjust its development process, in light of what we know about joint senior leader selection, to increase competitiveness for joint senior leader positions.

Importantly, while the interviewees were aware that the scope of the study was limited to the Air Force's officer corps, they tended to view that officer corps through their own cultural lenses. Both inside and outside the service, there seemed to be a tendency to view the USAF as a relatively cohesive whole in the context of joint senior leader development. There were few references to the need to focus on grooming any one type of Air Force officer for joint senior command. That said, many of the observations about Air Force officers seemed to imply a focus on the USAF's rated community, most likely because they tend to serve in command positions. In 2016, for example, 25 of 37 Air Force generals promoted were pilots.[45] While the reader should take into consideration this tendency to focus on pilots, this study makes no assumptions about whether rated officers are best positioned for senior joint positions. In fact, it takes the broader view that the Air Force should carefully consider who it grooms for joint command in light of the requirements for joint senior leader selection, as discussed in the recommendations at the conclusion of this report.

[45] Rep. Mike Rogers, "Remarks to 2017 Space Symposium," April 4, 2017.

The Air Force Approach to Joint Senior Leader Development

There are both subjective and objective elements that shape the Air Force's development process for joint senior leaders. In general, the process for promoting Air Force generals is relatively opaque, managed out of the General Officer Matters Office. After an officer reaches two stars, there are no more formal promotion boards or officer performance reports. Position descriptions explicitly defining qualifications for these positions are either dated or nonexistent.[46] This creates a space for cultural norms within the service to pervade the general officer promotion process. This section will first describe subjective factors that influence senior joint leader development, and it will then turn to a discussion of the objective factors that drive the process.

Subjective Factors That Influence Air Force Joint Senior Leader Development

Subjective factors that may influence USAF competitiveness include (1) potential to favor candidates with overly specific service backgrounds, or "ducks picking ducks," and (2) a tendency to withhold top candidates for USAF positions and/or put forth underqualified candidates for senior joint positions.

Ducks Picking Ducks

Within the military services, including the USAF, the lack of standardized criteria for the selection of three- and four-star generals leads to the phenomenon of "ducks picking ducks."[47] Senior uniformed leaders might identify officers for three- or four-star slots based on their shared experience in the same military occupation. In the USAF, for example, there has historically been a tendency for pilots to be promoted at rates much higher than officers in other specialties, although that trend has become less prominent in recent years.[48]

One significant reason for this phenomenon is that leaders like to pick people who look like them and who share common experiences.[49] In other words, leaders know that they had to develop certain skills to reach their position in the service, and they tend to assume that those skills are most likely to guarantee the success of subordinates. In the USAF, this has often manifested itself in the tendency for the specialty dominating the service to protect the promotion of its own to the most senior ranks of Air Force leadership, with little emphasis on (1) the promotion of other career fields or (2) the promotion of Air Force officers within the joint

---

[47] ADM Michael Mullen, "Joint Chiefs of Staff Speech to the Naval War College," Newport, R.I.: Naval War College, January 8, 2010.

[48] Jeffrey Smith, *Tomorrow's Air Force: Tracing the Past, Shaping the Future*, Bloomington, Ind.: Indiana University Press, 2014, Figures 11.1, 11.2, and 11.3, pp. 150–153.

[49] Mullen, 2010; and David Barno, Nora Bensahel, Katherine Kidder, and Kelley Sayler, "Building Better Generals," report release event held by the Center for New American Security, Washington, D.C., October 28, 2013, p. 27.

community. In the Air Force's early days, it was the bomber pilot leadership who reflected this tendency. For example, Gen Curtis LeMay, in his position as USAF Vice Chief of Staff, sought to maintain the bomber generals' grip on leadership by continuing spot promotions at Strategic Air Command (SAC) after World War II. Air Force civilian leaders actually returned brigadier general promotion lists to him on the grounds that he was promoting too many bomber pilots.[50] General LeMay also believed in grooming bomber pilots in a parochial manner, keeping them at SAC for their whole career because he considered SAC the heart of airpower and, therefore, able to provide all the education an officer needed.[51]

By the time the GNA was enacted in 1986, the bomber generals' dominance had given way to the rise of the fighter generals, but the parochial tendency for leaders to groom subordinates from their own tribe on an Air Force–centric promotion path continued. Gen Merrill McPeak, the chief of staff of the U.S. Air Force (CSAF) from 1990 to 1994, openly favored fighter pilots for promotion over all others and focused their development in Air Force–centric jobs.[52] "The service's purpose is to generate combat capability that protects the country, and not necessarily to provide equal career opportunities for those who fly heavies [transport aircraft] or, heaven forbid, don't wear wings at all," he said in 1991.[53] In a move that many perceived as an effort to build fighter pilots' leadership credentials for senior Air Force command, General McPeak created composite wings of fighters, bombers, and transport aircraft. Normally, fighter pilots led these wings, so opportunities for other airmen to develop leadership skills at the wing level were limited.[54]

The problem with the "ducks picking ducks" approach, which encourages the development of a select group of similar officers for Air Force command, is that it does not foster the development of officers with the breadth of experience necessary for joint jobs. Some argue that the USAF of today still suffers from the "ducks picking ducks" problem. The USAF, they say, may be grooming general officers with a parochial airpower mentality at the expense of the broad, interagency mindset required of senior joint officers. One retired Army general involved in senior joint leader selection said that he sees USAF culture promoting "a pilot-centric force in such a way that it is less conducive to leading large organizations." He added, "In the Air Force, it's you and your plane before it's you and your people." One active-duty general officer involved in joint senior leader selection echoed that sentiment, noting that, in his view, the USAF sees air component commander as the penultimate job, followed by USAF chief of staff. But this

---

[50] Col Mike Worden, *Rise of the Fighter Generals: The Problem of Air Force Leadership*, Maxwell Air Force Base, Ala.: Air University Press, 1998, p. 81.

[51] Worden, 1998, p. 142.

[52] Caitlin Lee, "The Culture of US Air Force Innovation: A Historical Case Study of the Predator Program," doctoral dissertation, King's College, London, 2017, pp. 126–128.

[53] Quoted in Wm. Bruce Danskine, "Fall of the Fighter Generals: The Future of USAF Leadership," thesis, Maxwell Air Force Base, Ala.: School of Advanced Airpower Studies, June 2001.

[54] Smith, 2014, pp. 109–112.

narrow focus on air-centric skills to get to the top is not conducive to four-star promotion in today's world, let alone joint command, the active-duty general officer argued: "By definition, the USAF doesn't groom four-stars. We still think the penultimate job is CFACC [Combined Forces Air Component Commander, with the ultimate goal of becoming USAF chief of staff], but if you go up that path, your odds of becoming a four-star are actually slim."

The literature also suggests a USAF tendency to focus on grooming new generations of airmen with pilot-centric skills to the detriment of winning joint posts. "There's something about the [USAF] culture that identifies the Air Force with specific positions rather than joint command," former Secretary of Defense Dick Cheney said in a 2000 interview.[55] GEN John Shalikashvili (U.S. Army, Ret.) echoed the sentiment in an interview with the same author: "[F]ew [top notch] Air Force officers have served on [joint or regional] staffs; the 'hot shots' all went to Air Force Headquarters."[56] Some USAF officers themselves have echoed this view, worrying that the service prizes technical skill over strategic thinking and air staff experience above joint experience.[57]

Withholding Top Candidates and Putting Forward Underqualified Candidates

"Ducks picking ducks" seems to have fueled a perception among interviewees inside and outside the service that the USAF may be withholding its most promising officers for service positions. One retired Army general involved in senior joint leader selection said he believed that, even 30 years after the enactment of GNA, services tend to hold back their best officers to fill critical service positions first and joint ones second. One active-duty general involved in senior joint leader selection said that this was indeed still a problem in the USAF, which tends to send its best officers to the Air Staff rather than joint billets.

While interviewees worried that the "ducks picking ducks" phenomenon may be leading the USAF to withhold its best officers, they also noted that it could hurt the USAF's reputation in the joint community in a second way. Several interviewees noted that "ducks picking ducks" creates a perception that USAF officers lack sufficient joint experience and are therefore less competitive for senior joint positions. In particular, USAF officers may receive career training for USAF leadership positions at the expense of the kind of experience required to excel in senior joint billets. A retired Navy admiral who was involved in senior joint leader selection confirmed that the USAF is failing to put forth competitive candidates for Joint Staff positions, although he said he could not say whether this was because the USAF was holding back its best candidates or simply because the USAF did not have good candidates to begin with. "The USAF bench seemed thin," he said. "It was not an uncommon refrain from a service chief to say 'I just

---

[55] Belote, 2000, p. 70.

[56] Belote, 2000, pp. 75–76.

[57] Stephen E. Wright, "Two Sides of the Coin: The Strategist and the Planner," in Richard Bailey and James Forsyth, eds., *Strategy: Context and Adaptation from Archidamus to Airpower*, Annapolis, Md.: Naval Institute Press, 2016, pp. 232–252.

don't have anybody who can play.' Well, what is going on in the Air Force that you can't compete here?"

There was a minority view, shared by two other retired Navy admirals involved in joint leader selection, that USAF candidates were generally competitive. "The Air Force has been very good about filling billets on the joint staff," said one of the admirals. "The quality was very good."

Objective Factors That Influence Air Force Joint Senior Leader Development

In addition to the subjective influences on USAF competitiveness for senior joint positions, there are also some objective ones. These factors can be adjusted through policy changes, but these changes would need to be underwritten by a cultural shift away from the service-centric approach to senior leader development discussed above toward a greater focus on the joint community. There are at least four objective factors that shape the USAF's competitiveness for senior joint positions. They include (1) the nature of joint experience, (2) the extent of joint experience, (3) the extent of strategic-level education, and (4) organizational capacity.

Nature of Joint Experience

The nature of joint experience, in terms of the types of exposure that airmen gain, is widely viewed as a key consideration in the selection of joint senior leaders. Officers need to make sure that they are getting exposed to the "right" kinds of joint experience. On one hand, Joint Staff jobs are important, not only to gain exposure to selection officials but also to ensure that the officer gains staff experience and understands Washington politics. On the other hand, deployed experience that significantly exposes airmen to other branches of service could also increase airmen's competitiveness. Interviewees cited the examples of General Breedlove, who served as an air liaison officer, working closely with the Army, early in his career, and Lt Gen Steven Shepro, who gained significant experience working closely with ground forces as a commander in both Iraq and Afghanistan. Interviewees cautioned, however, that while high-quality joint experiences downrange are valuable for gaining exposure, "box-checking" tours that double-count as a deployment and a joint tour are not particularly useful. "The USAF's stock is falling" in the joint world because too many airmen are trying to get a "two-fer" in this way, according to one active-duty general involved in senior joint leader selection.

Extent of Joint Experience

The extent of an officer's joint experience also plays a major role in determining how far he or she will climb up the joint community ladder. As mentioned above, certain Joint Staff positions (DJS, DJ3, DJ5, and DJ8), as well JTF commander positions, are seen as important grooming assignments. But winning those positions requires significant joint experience, which needs to be developed through exposure to even more junior joint positions earlier in an airman's career.[58] One retired USAF general who served in a senior joint position argued that airmen should start gaining "meaningful joint experience" at the O-6 level in J3 and J5 positions at the combatant commands or the Joint Staff. Another active-duty USAF general involved in senior joint leader selection suggested that joint experience should begin even earlier, noting that an airman who served on the joint staff as a major or a lieutenant colonel has an "astronomically" better chance of getting a key joint position when returning as a one-star.

USAF officers may not be spending enough time in junior joint assignments. In a 2009 study, a RAND researcher found that, in 2007, only four out of 16 brigadier general selectees had served on the Joint Staff, and only one had completed an assignment in OSD.[59] More research would need to be done to confirm whether this is still the case.

Extent of Strategic-Level Education

Most interviewees felt that USAF officers were able to take a holistic view of strategy in joint positions, stepping outside of the airpower perspective. But one retired Army general involved in senior joint leader selection said that he often had to lecture his air component commanders about the need to focus on the task at hand—conducting close air support operations in support of ground forces—rather than casting operations in terms of strategic bombing doctrine, a central tenet of airpower theory.

In any event, some interviewees felt that all of the services could do a better job helping officers learn to think strategically. Another retired Army general involved in joint senior leader selection noted that officers spend their early careers in a tactical mindset and then are expected to immediately take a strategic view once they become generals. He felt that all the services could do more to help with this transition. Joint academic education is crucial to help officers broaden their views so they can "think and act like Washington professionals," said one active-duty general involved in joint senior leader selection. Combatant commanders, in particular, need to take a strategic view so that they can effectively deal with interagency players in Washington, including the State Department and the intelligence communities, added another retired Army general involved in joint senior leader selection.

Organizational Capacity

Several interviewees noted that the USAF is not well structured to stand up JTFs, but they had mixed views about whether the USAF could or should make changes. The main obstacles, according to interviewees, seem to be the tendency for geographic combatant commanders, particularly Army generals, to pick commanders from their home service because of (1) service bias and (2) the Army's JTF-friendly organizational design at the corps and division levels.

---

[58] Goldfein, October 2016b.

[59] Robbert, 2009, slide 8.

The USAF is considering several options to optimize the 9th Air Force for JTF operations. One of those options is to create a JTF- ready organization similar to the Army's. But one active-duty general warned that this construct has the potential to waste combat power, since the units within the 9th Air Force would be mostly training to deploy as a JTF, rather than focusing on the in-garrison mission. Other options would involve streamlining the 9th Air Force and other numbered air forces (NAFs) for JTFs to lesser degrees. But these approaches might not fully address the combatant commanders' concerns about the organizational capabilities of the USAF's JTF-capable units.

Ultimately, whether it makes sense to reorganize the USAF to better support JTFs may depend on the USAF's goals. If the USAF wants to be ready to support a broader variety of JTF's, including ground-centric ones, such reorganization may be a necessary precondition, given the tendency of geographic combatant commanders to turn to organizations that are tailor-made to support ground-centric JTFs. That said, if the USAF did reorganize to support a broader variety of JTFs, there would certainly be no guarantees that combatant commanders would be more likely to pick the USAF to stand up a JTF, especially in light of their past reluctance to do so.

If, in contrast, the USAF is most interested in leading air-centric JTFs, it may still decide that change is required to develop organizations with organic JTF capabilities to appeal to geographic combatant commanders. Then again, the service may ultimately conclude that it is not worth the time and effort to reorganize. It may be the case that the USAF views air operations centers as adequate to run air-centric JTFs, particularly since they have expanded their domain focus to become Multi-Domain Operations Centers (MDOCs).

The Joint Senior Leader Selection Process

Having discussed the factors that drive Air Force joint senior leader development, we will now turn to an examination of the various subjective and objective factors that may influence the joint senior leader selection process. This process typically involves the CJCS, VCJCS, and the services, which agree on candidates who are then recommended to the Secretary of Defense. The Secretary of Defense normally accepts the recommendations and forwards them to the President, who nominates the candidate to a given post, pending Senate approval. That said, the Secretary of Defense can, and does, send back recommendations to ask for new candidates if he does not like the pool. Notably, Secretary of Defense Donald Rumsfeld played a far more activist role in selecting candidates for senior joint posts during his tenure from 2001 to 2006. But interviewees told us that, generally, the process has rested largely in the hands of the CJCS, VCJCS, and the service chiefs.

Subjective Factors That Influence the Joint Senior Leader Selection Process

The joint senior leader selection process has changed over time, but the one constant has been that it has been largely subjective. As mentioned above, position descriptions for joint senior leader positions are often outdated and vague. DoD does not conduct performance reviews for general officers beyond the rank of two stars. But even if there were a set of enduring, consistent, objective selection criteria for joint senior leaders, selectors would still bring their own perceptions and biases into the selection process.[60] In fact, they would be negligent if they did *not* do so. The current business literature suggests that while objective measurements of competency are important for successfully choosing high performers, subjective views of candidates—attained through in-depth interviews and reference checks—also must be considered to gauge notoriously intangible qualities, such as leadership potential, in future positions.[61] While this subjectivity has an upside in the sense that it takes into account intangible qualities at both the individual candidate level and the service level, it also allows for the selection officials' own biases to potentially influence decisions about senior joint leader selection. Our research found that there are at least four subjective factors that weigh heavily on the senior joint leader selection process: (1) subjective leadership competency criteria, (2) service biases, (3) perceptions of military domain knowledge, and (4) interpersonal relations.

Subjective Leadership Competency Criteria

One major subjective factor that drives the joint senior leader selection process relates to competency criteria. Because position descriptions for joint positions are often vague and outdated, there has never been any objective baseline for senior civilian and JCS leadership to use to select leaders for senior joint positions. Many desired competencies seem to be drawn from broad and vague ideas about what makes a good military commander. Critical skills cited in the literature on military command include strategic thinking, political savvy, and interpersonal skills.[62] A U.S. Army survey asking general officers to select general officer leadership qualities at the division level unearthed a bevy of qualitative attributes that might come under consideration, from "influences others without relying on rank or position" to "demonstrates

---

[60] The bias is not necessarily part of a deliberate or conscious effort. See, for example, Stefanie K. Johnson, David R. Hekman, and Elsa T. Chan, "If There's Only One Woman in Your Candidate Pool, There Is Statistically No Chance She'll Get Hired," *Harvard Business Review*, April 26, 2016.

[61] On the difficulty of measuring potential, see Malcolm Gladwell, "Most Likely to Succeed," *New Yorker*, December 15, 2008. On the need for subjective selection criteria in addition to objective ones, see Claudio Fernández-Aráoz, Boris Groysberg, and Nitin Nohria, "How to Hang on to Your High Potentials," *Harvard Business Review*, October 2011; and Claudio Fernández-Aráoz, "21st Century Talent Spotting," *Harvard Business Review*, June 2014.

[62] Belote, 2000, pp. 7–12, 62–63.

intellectual curiosity/life long learner."[63] Pentagon leaders involved in the selection of senior joint leaders have occasionally tried to formalize these criteria. Secretary Rumsfeld briefly attempted to codify selection criteria, and Gen Peter Pace provided a general description of the need for joint officers to be "strategically minded" and to possess "critical thinking skills" in his 2005 CJCS Vision for Joint Officer Development.[64] But these efforts have not provided a clear-cut and consistent rubric for candidate selection and nomination. In the end, selection officials rely on their judgment, "looking for someone who fits the right mold," said one retired U.S. Army general involved in senior joint leader selection. "A lot has to do with personality—how they fit in to the team—that has an impact."

Service Biases

Biases both for and against various services at the OSD level may also play a role in joint senior leader selection. In 2008, for example, the now-defunct U.S. Joint Forces Command issued a report on joint operations best practices. "We find that the services . . . have unique skill sets in terms of being more suited for 'filling' the different staff principle positions," the report said. "For example, the USA [U.S. Army] and the USMC [U.S. Marine Corps] cultures and assignments seem to produce effective CoS (chiefs of staff) and J3s, and USAF and USN [U.S. Navy] have unique J2 (intelligence), J5 and J6 attributes (command, control, communications and computers/cyber)."[65] No empirical research was offered to support these statements.

Service biases might also come into play if a senior leader from one service is in charge of selecting a leader for another senior joint position. This scenario is most relevant to the selection of JTF commanders. In theory, the combatant commander will stand up a JTF from a service responsible for the preponderance of forces in a given region,[66] but in practice, the decision is colored by a variety of factors, including organizational fit and service skill sets, but also, potentially, service bias.[67] Geographic combatant commanders may, for example, tend to select JTF commanders from their home service, as has been the case with CJTF-OIR, USCENTCOM's JTF charged with defeating the Islamic State. The commander of USCENTCOM has been a U.S. Army general since 2013 (GEN Lloyd Austin, followed by GEN Joseph Votel), and the last three commanders of CJTF-OIR, established in 2014, have also been

---

[63] Survey copy courtesy of study lead R. Craig Bullis, U.S. Army War College lead researcher. Results were not available as of this study's completion (R. Craig Bullis, *Survey of Leadership Competencies*, U.S. Army War College, forthcoming).

[64] Joint Chiefs of Staff, *CJCS Vision for Joint Development*, November 2005.

[65] Gen Gary Luck (Ret.), Col Mike Findlay, and the JWFC Joint Training Division, *Joint Operations: Insights and Best Practices*, 2nd ed., July 2008, p. 25.

[66] Spirtas, Young, and Zimmerman, 2009, p. 3.

[67] Spirtas, Young, and Zimmerman, 2009, pp. 16–17.

Army generals (LTG Stephen J. Townsend is the current commander; he was preceded by LTG Sean MacFarland and LTG James Terry).[68]

In theory, a combatant commander could select a JTF commander from another service. But one retired Army general involved in joint senior leader selection said that an airman competing for a JTF commander position might face skepticism from both the combatant commander and Army personnel on his JTF staff. This would be partly because of a genuine concern that the USAF tilts toward training airmen rather than "joint" officers equipped to command joint operations (discussed more later), but also, he admitted, because of a service bias within the Army against having airmen command soldiers. "It would be a tough sell and there would be a lot of reticence to do that culturally," he said. "But in the long term, that is where we have to get. You will get pushback in the Army, but the people you want in those positions are the best athletes."

Perceptions of Military Domain Knowledge

Another subjective factor that affects decisionmaking for senior joint billets relates to perceptions of military domain knowledge. There are at least two ways in which perceptions of military domain knowledge influence commander selection for these organizations. First, selection officials consider candidates' knowledge of a given geographic domain, such as Europe or Africa. Second, selection officials consider whether the candidate has an ability to work across warfighting domains. Interviewees said that, generally, knowledge of the geographic domain is more important than cross-domain knowledge.

Interviewees all agreed that it was critical to have geographic domain knowledge, which could be built up through a series of assignments in a given theater. "Most important is geopolitical domain knowledge," said one retired Army general involved in joint senior leader selection. "The ability to get up to the strategic level, which is what we are focusing on . . . that is the most important thing."

But, as mentioned earlier, several interviewees also cautioned that politics also plays a major role in determining who wins top positions at geographic commands. For example, the Navy has historically held the commander position at USPACOM by citing its long history of operations in the Pacific and its view that naval forces are most critical to warfighting in the region. When necessary, the Navy draws on outside support to bolster its case. Senator John McCain, himself a naval aviator and his father a former USPACOM commander, has been a strong advocate of the Navy's continued dominance in the Pacific AOR.[69] Deceased Hawaii Senator Daniel Inouye also was predisposed to favor a naval officer for the position.[70] As a result, even if a service grooms a

---

[68] U.S. Department of Defense, "Townsend Takes Command of Operation Inherent Resolve," August 21, 2016; LTG James L. Terry, "US Army Central: Operating in a Complex World," U.S. Army, October 1, 2015.

[69] Grant, 2008, p. 48.

[70] Gregg K. Kakesako, "General Pulls Plug on Camp Smith Job," *Honolulu Star Bulletin*, October 7, 2004.

general officer with significant geographic domain expertise, biases such as that in favor of the Navy in the Pacific may prove to be the overriding consideration.

Selection officials also viewed cross-domain knowledge as an important trait of candidates for senior joint billets. One retired Army general involved in joint senior leader selection said that he worried that USAF officers were not adequately trained in joint warfighting and generally lacked experience on the battlefield. He questioned whether there are many "warfighters" in the USAF, other than perhaps former USAF Chief of Staff Michael Moseley and the current chief, Gen David Goldfein, who was shot down over Serbia. He seemed to equate the term *warfighter* with the possession of empathy for ground forces and the establishment of trust that can only be gained by forming relationships with ground forces in battle. "It's hard to find those kinds of guys in the Air Force," he said. "I would feel better if I knew who the warfighters are." A former Secretary of Defense involved in joint leader selection added that USAF officers' perceived lack of connection with ground forces—due, in part, to the nature of the air medium—could make it difficult for airmen to think through the operational risks of using ground forces.

Interpersonal Relations

Several interviewees emphasized that building trust with other services and outside actors over the course of one's career, mainly through joint assignments, is critical for promotion to the highest levels of joint command. These positions need to be highly visible to the most-senior civilian leaders. One active-duty general officer involved in joint senior leader selection noted that it is particularly important to get airmen into joint billets that "matter"—the DJ3 or the DJ5. "If you are an operator sitting in a DJ1 or DJ2 job, that is not helpful," he said. One example of effectively creating opportunities to develop relationships with senior people is the career of now-retired USAF Gen Philip Breedlove, who, in his positions as Vice Director for Strategic Plans and Policy (J5) on the Joint Staff from 2006 to 2008 and as USAF Vice Chief of Staff from January 2011 to July 2012, was able to develop a rapport with senior civilian and military leadership. Recognizing that these relationships put him a good position to earn a joint billet, Secretary of Defense Leon Panetta nominated General Breedlove as commander of U.S. Air Forces Europe, knowing that this assignment would make him a strong contender to become Supreme Allied Commander Europe/USEUCOM commander.[71] He later was selected and served in this position from May 2013 to July 2016.

Objective Factors That Influence Senior Joint Leader Selection

While subjectivity inevitably dominates the selection process, over the years, selection officials have engaged in four different practices that have inserted objectivity into the selection of joint senior leaders and JTFs. These include (1) considerations of diversity, (2) considerations

[71] However, General Breedlove's nomination occurred only after the first nominee, Gen. John Allen (USMC), declined to go through the nomination process and chose to retire (Jennifer H. Svan, "Gen. Breedlove nominated to head US European Command," *Stars and Stripes*, March 28, 2013).

of "fair share" representation among the services, (3) senior defense civilian activism, and (4) an evaluation of a service's organizational fit. While these are all important considerations in the selection process, the consensus of interviewees was that the first two, diversity and "fair share" representation among the services, are secondary concerns. Similarly, senior defense civilian activism, in which the Secretary of Defense steps in to provide an outsider perspective, is not typically a primary driver in the selection process. Notably, organizational capabilities—the extent to which a service is appropriately organized to fulfill a given mission—could be a decisive factor in standing up a JTF.

Diversity

Selection officials have taken into consideration certain objective, highly tangible factors, such as race or gender, as they develop a list of potential candidates for a position. This has been an important priority inside the Pentagon since the Vietnam War, when the small number of minority officers contributed to a perception of systemic racial discrimination, harming morale and heightening tension.[72] Between 1967 and 1991, the Pentagon almost quadrupled the number of minority officers, and the proportion of female officers grew ninefold.[73] From 1986 to 2006, minority officer representation above the rank of O-7 increased 9 percent.[74] That said, the total number of minority officers above the rank of O-7 remains small, with African Americans and Hispanic Americans representing only 5 percent of general officers above the rank of O-7 in 2006.[75] Interviewees involved in joint senior leader selection said that while diversity is not generally a deciding factor in selection, it does help to shape the candidate pool.

"Fair Share" Representation Among the Services

Selection officials have taken into account the concept of each service having a "fair share" of senior joint positions. Similar to diversity, equal representation among the services was not a primary consideration, but several interviewees who have been involved in the senior joint leader selection process said that service representation is a consideration. "We always tried to get a balance among the COCOMs [combatant commands] and the joint staff," said one retired Navy admiral involved in senior joint leader selection. "In my mind, I had a list of which services had which jobs, but it was just a filter." Similar to diversity, "fair share" representation was not a primary consideration, but it could be a tiebreaker.

---

[72] Nelson Lim, Jefferson P. Marquis, Kimberly Curry Hall, David Schulker, and Xiaohui Zhuo, *Officer Classification and the Future of Diversity Among Senior Military Leaders: A Case Study of the Army ROTC*, Santa Monica, Calif.: RAND Corporation, TR-731-OSD, 2009.

[73] Susan D. Hosek, Peter Tiemeyer, M. Rebecca Kilburn, Debra A. Strong, Selika Ducksworth, and Reginald Ray, *Minority and Gender Differences in Officer Career Progression*, Santa Monica, Calif.: RAND Corporation, MR-1184-OSD, 2001, p. xiii.

[74] Lim et al., 2009, p. 1.

[75] Lim et al., 2009, p. 2.

Senior Defense Civilian Activism

Senior defense civilians have periodically intervened in the substantive decisions surrounding joint leader selection, injecting a fresh perspective into the process that potentially mitigates other selection biases. Historically, the Secretary of Defense has reserved the right to send back recommendations for senior joint positions if he did not like the candidates. During a unique period in the history of senior joint leader selection, Secretary of Defense Donald Rumsfeld took an even more activist approach. He was concerned that the services were nominating candidates for senior joint positions based on cronyism—the "tendency . . . to pick those you know and to overlook better, lesser-known candidates"—rather than based on a thorough consideration, albeit an inevitably biased one, of the candidates' qualifications.[76] He sought to remedy that problem by playing an activist role as an outsider defense civilian willing to reach into the services to identify his own candidates for three- and four-star billets.[77]

After Secretary Rumsfeld left office, interviewees involved in senior joint leader selection told us that the authority to nominate candidates for senior joint positions fell back to the CJCS and the services. But more recently, there are indications that the current Secretary of Defense, James Mattis, may be interested in adopting some aspects of Secretary Rumsfeld's approach.[78] If the Secretary of Defense reemerges as a highly active player in joint senior leader selection, then understanding how senior defense civilians think about joint senior leader selection and the Air Force's competitiveness for these positions will become particularly important.

Organizational Considerations for the Establishment of JTFs

Selection officials have focused on organizational considerations in the specific case of JTFs. As mentioned earlier, some interviewees indicated that commanders tended to show bias toward their home service in the selection of JTF commanders, and JTF commanders, in turn, tended to prefer JTF staffs established from their home unit. But several interviewees also cited organizational design as a more objective reason that drives decisionmaking about the service responsible for a JTF.

The organization selected for the JTF must possess all the relevant deployable components, from communications to aviation and logistics. The U.S. Army's division- and corps-level units have historically dominated JTFs. This may be, at least in part, because of the desire on the part of the combatant commander standing up the JTF to choose a service organization that seems to objectively meet the missions required of the JTF. As noted earlier, the last three commanders of CJTF-OIR have come from the Army. While this could be, in part, because of the geographic

---

[76] Andrew Hoehn, Albert A. Robbert, and Margaret C. Harrell, *Succession Management for Senior Military Positions: The Rumsfeld Model for Secretary of Defense Involvement*, Santa Monica, Calif.: RAND Corporation, MG-1081-RC, 2011, pp. xiii–xiv and p. 18.

[77] Hoehn, Robbert, and Harrell, 2011, p. xiii.

[78] Secretary Mattis recently requested a copy of the RAND study on Secretary Rumsfeld (Hoehn, Robbert, and Harrell, 2011).

commanders' preference for a soldier to run the JTF, there may also be a sense that an Army general can most quickly and efficiently tap the home organization to serve as the JTF headquarters. In the case of the CJTF-OIR, the last three commanders all brought their home organizations with them. Interviewees, including an active-duty general officer involved in senior joint leader selection, said that the Army's corps- and division-level units possess the organic capability to readily stand up a JTF for cross-domain operations. The USAF's corresponding organizations—the numbered air forces—do not currently possess those same organic capabilities.

5. Findings and Recommendations

The USAF should worry about its representation in the most critical joint senior leader positions, according to a review of the literature and interviews with retired senior Air Force, Army, Navy, and OSD leaders. According to interviewees, the most critical positions are CJCS, VCJCS, DJS, DJ3, DJ5, DJ8, Joint Staff jobs, high-profile JTF commands, and combatant commands, particularly USCENTCOM, USPACOM, and USEUCOM. Interestingly, views of the importance of USSOCOM were more mixed, despite that command's highly prominent role in both warfighting and strategymaking on a global scale.

The USAF may be less competitive for senior joint positions for both subjective and objective reasons, according to the interviewees. Most notably, in terms of subjective considerations, interviewees perceived that the Air Force may have a cultural tendency to focus on grooming its rated force for top positions inside the USAF rather than systematically cultivating qualified officers for joint assignments. Objective considerations that may be reducing competitiveness include a potential shortfall in the quality of joint experience in terms of both Washington staff work and cross-domain exposure, a lack of joint experience early in airmen's careers, a lack of emphasis on strategic-level education focused on interagency cooperation and geographic expertise, and an inadequate organizational structure to support the establishment of JTFs.

To adopt meaningful reforms that address these shortfalls, the USAF needs to consider whether it is willing to undertake a cultural transformation. Any potential reforms of the objective factors that are seen to be hurting the USAF's competitiveness would have to be underwritten by a fundamental cultural shift. In today's Air Force, there is a perception, among interviewees both inside and outside the service, that the most talented airmen are rated officers who are groomed for senior positions *within* the Air Force. This is an obstacle to senior joint leader development both because the "right" kinds of talent for joint leadership may be underdeveloped and because the Air Force's best officers may be held back for Air Force assignments as opposed to joint ones. The USAF has to consider whether it is willing to take on reforms that will break through this cultural barrier, elevating the importance of senior joint command over senior Air Force command, if it is to meaningfully reform senior joint leader development.

If the USAF decides to embrace meaningful reform, it will need to (1) openly examine and acknowledge its values and priorities in regard to senior leader development and (2) use those values and priorities as a basis to make conscious decisions about where to invest time and resources in joint senior leader development while acknowledging the corresponding trade-offs. There are at least three major potential downsides to pursuing reforms related to joint senior leader development. First, to some extent, the senior joint jobs that are most important for the nation change over time as the strategic environment evolves. Second, a wide variety of external factors beyond the USAF's control impinge on the decisionmaking process that surrounds senior joint leadership selection. Third, to the extent that the USAF can influence its competitiveness for the most-critical senior joint billets, there is a broader question that the USAF needs to consider about whether the Air Force is willing to take on some risk in the development of its officers for senior Air Force positions in order to support enhanced joint senior leader development.

The following recommendations aim to help the USAF take a more purposeful approach to senior joint leader development. Interviewees worried that the lack of a well-understood, deliberate process for grooming joint senior leaders has allowed the USAF's historical "ducks picking ducks" culture to influence senior leader development. The first three recommendations below aim to create a more deliberate process for developing senior joint leaders in which the USAF cultural norms still play a role but are tempered by other considerations driving senior leader development. The last four recommendations suggest specific reforms, based on input from our interviewees, to improve USAF competitiveness for senior joint positions.

Decide Which Senior Joint Positions Are Most Important

Interviewees inside and outside the USAF generally agreed on the senior joint positions that were most important, defined in terms of those that are most crucial to the nation's strategymaking and warfighting capacity. But even within that subset, there was a hierarchy of positions based on perceived contributions to strategy and warfighting. There was a consensus that USCENTCOM, USEUCOM, and USPACOM were the most important geographic combatant commands, for example.

The USAF may want to have its own internal discussion among senior leadership about exactly which senior joint positions it considers most critical, and why. There would be two benefits to specifically identifying the Air Force's priorities for senior joint leadership. First, the Air Force could tailor joint senior leader development toward those positions, focusing more, for example, on gaining geographic expertise through education and joint tours. Second, and most importantly, making an explicit decision about which positions are most critical would give the Air Force the opportunity to think about whether it is willing to accept the risks involved in seriously competing for those positions. A first-order question might be: Is the nation's warfighting capacity at risk because airmen are not leading certain joint commands? If the answer to the question is "yes," is the Air Force willing to prepare some of its most promising officers for those positions, even with no guarantees of winning the position? What if gaining representation in that critical position meant giving up warfighting capacity in one area to gain it in another? After doing this analysis, the Air Force might conclude that improving joint representation in one area might not be worth the risks.

In the case of geographic combatant commands, for example, the Air Force would have to accept the risk of training officers to command in a specific geographic area, such as USCENTCOM, only to find that the command has become a backwater by the time that officer reaches three stars. Also, the Air Force would have to accept the risk of building a rated officer to be a geographic combatant commander, with all the time that might involve, only to find out that the vagaries of the senior joint leader selection process outlined in this report prevent that officer's selection. Finally, the Air Force would have to accept the risk that selection for a geographic command like USCENTCOM might detract from Air Force representation somewhere else, based on the selection officials' concerns about "fair share" representation. So, for example, selection for a USCENTCOM command position might mean losing the Air Force's historic lock on USTRANSCOM.

Select the Officer Candidate Pool

After deciding which senior joint positions are most critical to the Air Force, the service might then want to think about which officers are best suited to be groomed for those positions. The service may want to focus on a specific officer pool for three reasons.

First, reform of senior joint leader development may require a greater commitment to joint education and training. The time and money required to effectively enhance development of all Air Force officers for joint positions might exceed the time and resources available. As a result, the reform effort could end up being spread thinly and ineffectively across the officer corps.

Second, tailoring the officer pool for senior joint development also gives the Air Force an opportunity to think about which types of officers might be most competitive for the senior joint positions that the service sees as most critical. Is one particular Air Force specialty code better-suited for a track to become geographic combatant commander than the others? Or are there Air Force officers with certain types of experiences, such as leading large units, that might be best positioned for joint senior leader development?

Third, focusing on a specific pool of officers may help the Air Force think about the trade-offs associated with enhanced senior joint development. Most importantly, officers targeted for senior joint leader development might miss certain service-centric experiences that would have all but guaranteed their chances to rise through the ranks in Air Force command. The career path for senior joint leadership might still provide these officers excellent opportunities in the USAF, but that path might look somewhat different than if they were being strictly groomed for a top USAF post. Furthermore, there are no guarantees that these officers would be selected for sought-after joint posts. The USAF would need to consider whether it could afford to potentially sacrifice some of its top talent in the name of winning more senior joint representation.

Build a Succession Plan and Institutionalize General Officer Development Practices

Once the Air Force has decided which senior joint positions matter most and whom to groom for them, it may want to develop a succession plan for senior joint development. This would help to mitigate some of the problems that interviewees identified with Air Force leadership development, including the problems of putting forward underqualified candidates and withholding candidates for senior positions within the Air Force. The succession plan would ensure that candidates for senior joint positions pursue well-understood career pathways that aim to balance Air Force and joint experience. The succession plan might also involve looking across all the military services to see who the up-and-coming joint senior leaders are. More-junior Air Force officers who have served with those individuals could be placed in strong positions to compete for critical joint senior leader development positions under those rising stars.

To ensure that this succession plan is enacted, the Air Force might also want to consider adopting a standard set of well-understood senior leader development practices. Commanders would be briefed on the succession plan and would understand how and why certain officers were being groomed for certain positions. This would require a top-down management style that would take away some of the authority of Air Force commanders to pick their own people, but it would ensure that candidates selected for senior joint leader positions received the experience they needed to succeed.

Increase the Quality of Joint Experience

In terms of increasing the quality of joint experience, the USAF should consider three steps.

1. It should ensure that it is sending its officers to high-visibility positions to allow for networking opportunities and to expose airmen to Washington staff work.

2. It needs to ensure that airmen serve in sufficient joint warfighting tours that increase cross-domain understanding, thereby increasing airmen's credibility to lead joint organizations.

3. The USAF should consider concentrating airmen's careers in a given geographic area of responsibility, providing an opportunity to gain subject-matter expertise in a given geographic domain.

Increase the Extent of Joint Experience

In terms of the extent of joint experience, the Air Force should consider starting joint assignments earlier for officers who are targeted for senior joint positions. Interviewees recommended that these officers gain joint experience as early as the rank of major or lieutenant colonel because airmen who had a junior joint tour were far more likely to be chosen for the more senior joint assignments.

The downside of ensuring that airmen receive quality joint experience is that it takes time. Airmen today may be opting for "box-checking" joint tours that double-count as a deployment and a joint assignment but do not provide sufficient exposure for promotion within the joint community. The USAF has to decide whether it is willing to purposefully eliminate box-checking and devote more time within at least a certain number of airmen's careers to gain joint experience. There are two ways this could be accomplished, but they require making trade-offs.

First, the Air Force could provide an opportunity for at least some officers to focus on joint assignments at the expense of time spent in service-centric assignments. For example, an airman on the joint track might skip a USAF-centric command assignment (such as group command) and instead use that time to gain joint experience.

Second, the USAF might consider extending the joint and strategic experience window by creating more flexibility around the "24-year pole," the time frame around which officers are generally promoted to brigadier general.

Reconsider the Nature and Extent of Joint and Strategic Education

The USAF should consider educational reforms to enhance airmen's ability to think strategically. In particular, interviewees suggested that more attention should be paid to understanding the interagency process. They also suggested that officers might benefit from studying the specific geographic regions where they hope to do future joint assignments.

The main cost of this approach may be that airmen have to sacrifice some time that would have been spent learning about airpower theory, doctrine, or other types of USAF-specific knowledge. It also will require a more demanding educational experience, at least for those selected officers chosen to compete on the joint track. Finally, as is the case with joint experience reforms, there would be no guarantees that this educational strategy would win the USAF more critical senior joint assignments, and, in the case of geographic specialization, there are no guarantees that a given area specialty would still be relevant by the time that airman was ready for a senior joint position in his or her geographic domain of expertise.

Consider Organizational Reforms for JTFs

Any organizational reform to accommodate the stand-up of JTF capabilities should be viewed in the context of what the Air Force hopes to achieve. If the USAF is only interested in running air-centric JTF operations, it may be the case that the existing MDOC organization is optimal. If, however, the USAF desires to expand its capacity to manage cross-domain JTFs, it may make sense to reconsider JTF organization. Given the biases involved in the JTF stand-up process, the USAF would need to pursue the option to build a full JTF capability, along the lines of the U.S. Army's organic capabilities at the division and corps units, to have the best chance of being chosen to run cross-domain JTFs.

One downside of building this organic JTF capability, however, is that the new structure may not be enough to overcome geographic combatant commander biases, which might continue to prevent the USAF from managing a major cross-domain JTF. In addition, the USAF would need to consider whether it could afford to dedicate significant USAF base resources in garrison to a forward-deployed JTF. Although some current small and unique units are organized this way, this would be a significant departure from the USAF's current model, under which USAF resources focus on the in-garrison mission and airmen deploy as smaller units and augmentees to support a JTF.

Lead Joint Senior Leader Development Reform from the Top Down

Undertaking reform to foster senior joint leader development will require a cultural shift, at least to some degree, away from an Air Force–centric approach to a more joint outlook. Because senior leaders play a key role in promoting cultural change, the Air Force's decision to embed this reform effort at the CSAF level seems important. Without senior Air Force leaders' support, there is a risk that any pool of officer candidates singled out for senior joint leader development might be viewed as outsiders whose career development does not need to be taken as seriously as that of Air Force–centric officers. Even with senior Air Force leadership support, these joint-tracked officers may experience some pushback from within the Air Force, particularly among Air Force commanders who are asked to respect a joint senior leader succession plan that takes away some of their authority to choose subordinates. Given these considerations, careful stewardship of the joint senior leader development effort is needed at the very top of the Air Force to ensure that the effort is taken seriously and leads to a meaningful cultural shift in support of the joint community.

Bibliography

Anderegg, C. R., *Sierra Hotel: Flying Air Force Fighters in the Decade After Vietnam*, Washington, D.C.: Air Force History and Museums Program, 2001.

"Army Brass Losing Influence," Associated Press via Military.com, June 15, 2007. As of June 29, 2017: http://www.military.com/NewsContent/0,13319,139239,00.html

Barno, David, Nora Bensahel, Katherine Kidder, and Kelley Sayler, "Building Better Generals," report release event held by the Center for New American Security, Washington, D.C., October 28, 2013. As of November 3, 2016: https://www.cnas.org/events/building-better-generals

Belote, Howard D., Lt Col (USAF), *Once in a Blue Moon: Airmen in Theater Command—Lauris Norstad, Albrecht Kesselring, and Their Relevance to the Twenty-First Century Air Force*, Cadre Paper No. 7, Maxwell Air Force Base, Ala.: Air University Press, July 2000. As of November 3, 2016: http://www.au.af.mil/au/aupress/digital/pdf/paper/cp_0007_belote_once_in_blue_moon.pdf

Belote, Howard D., Lt Col (USAF), "Proconsuls, Pretenders or Professionals? The Political Role of Regional Combatant Commanders," in *Essays 2004: Chairman of the Joint Chiefs of Staff Strategy Essay Competition*, Washington, D.C.: National Defense University, 2004.

Blumenson, Martin, and James L. Stokesbury, *Masters of the Art of Command*, Boston, Mass.: Houghton Mifflin, 1975.

Bonds, Timothy, Myron Hura, and Thomas-Durell Young, *Enhancing Army Joint Force Headquarters Capabilities*, Santa Monica, Calif.: RAND Corporation, MG-675-1-A, 2010. As of November 3, 2016: https://www.rand.org/pubs/monographs/MG675-1.html

Bowden, Mark, "The Desert One Debacle," *The Atlantic*, May 2006. As of December 21, 2016: http://www.theatlantic.com/magazine/archive/2006/05/the-desert-one-debacle/304803

Bowie, Chris, AF/XPX, "Buddy Could You Spare Five Billion Dollars: Or, Why Being a Staff Puke Is a Noble Calling," presentation to LTG Duncan McNabb, January 31, 2003.

Builder, Carl, *Masks of War*, Baltimore, Md.: Johns Hopkins University Press, 1989.

Bullis, R. Craig, *Survey of Leadership Competencies*, U.S. Army War College, forthcoming.

Burks, William H., Maj (USAF), "Blue Moon Rising? Air Force Institutional Challenges to Producing Senior Joint Leaders," master's thesis, Fort Leavenworth, Kan.: School of Advanced Military Studies, U.S. Army Command and General Staff College, 2010. As of December 29, 2016:
http://www.dtic.mil/dtic/tr/fulltext/u2/a523051.pdf

Cahlink, George, "Procurement Scandal Cuts Short Air Force General's Quest for Command," *Government Executive*, November 5, 2004. As of January 3, 2017:
http://www.govexec.com/defense/2004/11/
procurement-scandal-cuts-short-air-force-generals-quest-for-command/17963/

Cancian, Mark, "We Need a Map for Goldwater-Nichols Reform So We Don't Get Lost," *War on the Rocks*, March 17, 2016. As of November 3, 2016:
http://warontherocks.com/2016/03/
we-need-a-map-for-goldwater-nichols-reform-so-we-dont-get-lost/

Carter, Ashton, "Remarks on Goldwater Nichols at 30: An Agenda for Updating," Center for Strategic and International Studies, April 5, 2016. As of November 3, 2016:
http://www.defense.gov/News/Speeches/Speech-View/Article/713736/
remarks-on-goldwater-nichols-at-30-an-agenda-for-updating-center-for-strategic

Cole, Ronald H., Walter S. Poole, James F. Schnable, Robert J. Watson, and Willard J. Webb, *A History of the Unified Command Plan, 1946–2012*, Washington, D.C.: Joint History Office, Office of the Joint Chiefs of Staff, 2013.

Combined Joint Task Force—Operation Inherent Resolve Public Affairs Release, "III Corps Assumes Operation Inherent Resolve Mission," *Fort Hood Sentinel*, September 24, 2015. As of December 28, 2016:
http://www.forthoodsentinel.com/news/
iii-corps-assumes-operation-inherent-resolve-mission/
article_cfb218bb-46a9-5234-8c32-e9461fd61c1d.html

Danskine, Wm. Bruce, "Fall of the Fighter Generals: The Future of USAF Leadership," thesis, Maxwell Air Force Base, Ala.: School of Advanced Airpower Studies, June 2001. As of April 25, 2017:
http://indianstrategicknowledgeonline.com/web/danskine.pdf

Deptula, David, Gen (USAF, Ret.), "Perspectives on Air Force Positions in Joint and COCOM Senior Leader Positions," presented at the Air Force Association Symposium, February 22, 2013, Orlando, Fla.

DoD—*See* U.S. Department of Defense.

Ehrhard, Thomas P., *An Air Force Strategy for the Long Haul*, Washington, D.C.: Center for Strategic and Budgetary Assessments, 2009. As of November 3, 2016: http://csbaonline.org/uploads/documents/2009.09.17-An-Air-Force-Strat.pdf

Eisenhower, Dwight D., *Crusade in Europe*, Baltimore: Johns Hopkins University Press, [1948] 1997.

Feickert, Andrew, *The Unified Command Plan and Combatant Commands: Background and Issues for Congress*, Washington, D.C.: Congressional Research Service, R42077, January 3, 2013. As of December 28, 2016: https://fas.org/sgp/crs/natsec/R42077.pdf

Fernández-Aráoz, Claudio, "21st Century Talent Spotting," *Harvard Business Review*, June 2014. As of February 2, 2017: https://hbr.org/2014/06/21st-century-talent-spotting

Fernández-Aráoz, Claudio, Boris Groysberg, and Nitin Nohria, "How to Hang on to Your High Potentials," *Harvard Business Review*, October 2011. As of February 2, 2017: https://hbr.org/2011/10/how-to-hang-on-to-your-high-potentials

Gellman, Barton, "Air Force Promotions Tainted," *Washington Post*, November 28, 1991. As of April 24, 2017: https://www.washingtonpost.com/archive/politics/1991/11/28/air-force-promotions-tainted/85325dd3-8704-42a5-a8c8-e414e4777088/?utm_term=.345c18e34d1d

Gladwell, Malcolm, "Most Likely to Succeed," *New Yorker*, December 15, 2008. As of February 2, 2017: http://www.newyorker.com/magazine/2008/12/15/most-likely-to-succeed-malcolm-gladwell

Goldfein, David S., Gen (USAF), remarks at the 2016 Air, Space and Cyber Conference, Gaylord Convention Center, Oxon Hill, Md., September 20, 2016a.

Goldfein, David S., Gen (USAF), *CSAF Focus Area: Strengthening Joint Leaders and Teams*, Washington, D.C.: U.S. Air Force, October 2016b. As of November 3, 2016: http://www.af.mil/Portals/1/documents/csaf/letters/16%2010%2013%20Focus%20Area%20II.pdf?ver=2016-10-13-105649-460×tamp=1476371621707

Goldwater-Nichols Department of Defense Reorganization Act of 1986 (Pub. L. 99-433), October 1, 1986. As of November 3, 2016: http://history.defense.gov/Portals/70/Documents/dod_reforms/Goldwater-NicholsDoDReordAct1986.pdf

Gordon, Michael R., "After Hard-Won Lessons, Army Doctrine Revised," *New York Times*, February 8, 2008. As of December 29, 2016: http://www.nytimes.com/2008/02/08/washington/08strategy.html

Gould, Joe, "As Congress Pushes Defense Department Reform, So Does Carter," *Federal Times*, December 14, 2015. As of November 3, 2016: http://www.federaltimes.com/story/government/management/2015/12/14/congress-pushes-defense-department-reform-so-does-carter/77311426/

Grant, Rebecca, "Why Airmen Don't Command," *Air Force Magazine*, March 2008, pp. 46–49. As of July 12, 2017: http://www.airforcemag.com/MagazineArchive/Pages/2008/March%202008/0308command.aspx

Halperin, Morton, Priscilla Clapp, and Arnold Kanter, *Bureaucratic Politics and Foreign Policy*, 2nd ed., Washington, D.C.: Brookings Institution Press, 2006.

Hamre, John, "Keep America's Top Military Officer Out of the Chain of Command," *Defense One*, March 15, 2016. As of November 3, 2016: http://www.defenseone.com/ideas/2016/03/keep-americas-top-military-officer-out-chain-command/126694

Hoehn, Andrew R., Adam Grissom, David A. Ochmanek, David A. Shlapak, and Alan J. Vick, *A New Division of Labor: Meeting America's Security Challenges Beyond Iraq*, Santa Monica, Calif.: RAND Corporation, MG-499-AF, 2007. As of November 3, 2016: http://www.rand.org/pubs/monographs/MG499.html

Hoehn, Andrew, Albert A. Robbert, and Margaret C. Harrell, *Succession Management for Senior Military Positions: The Rumsfeld Model for Secretary of Defense Involvement*, Santa Monica, Calif.: RAND Corporation, MG-1081-RC, 2011. As of June 29, 2017: https://www.rand.org/pubs/monographs/MG1081.html

Hoffman, Michael, "Obama Nominates First Airmen for JCS Leadership Position in 10 Years," Military.com, May 5, 2015. As of November 3, 2016: http://www.military.com/daily-news/2015/05/05/obama-nominates-first-airman-for-jcs-leadership-position-in-10.html

Hosek, Susan D., Peter Tiemeyer, M. Rebecca Kilburn, Debra A. Strong, Selika Ducksworth, and Reginald Ray, *Minority and Gender Differences in Officer Career Progression*, Santa Monica, Calif.: RAND Corporation, MR-1184-OSD, 2001. As of June 29, 2017: https://www.rand.org/pubs/monograph_reports/MR1184.html

Huntington, Samuel, "Defense Organization and Military Strategy," *National Affairs*, No. 75, Spring 1984, pp. 20–46. As of December 20, 2016: http://www.nationalaffairs.com/public_interest/detail/defense-organization-and-military-strategy

Johnson, David E., *Learning Large Lessons: The Evolving Roles of Ground Power and Air Power in the Post-Cold War Era*, Santa Monica, Calif.: RAND Corporation, MG-405-1-AF, 2007. As of November 3, 2016:
http://www.rand.org/pubs/monographs/MG405-1.html

Johnson, Stefanie K., David R. Hekman, and Elsa T. Chan, "If There's Only One Woman in Your Candidate Pool, There Is Statistically No Chance She'll Get Hired," *Harvard Business Review*, April 26, 2016. As of February 3, 2017:
https://hbr.org/2016/04/
if-theres-only-one-woman-in-your-candidate-pool-theres-statistically-no-chance-shell-be-hired

Joint Chiefs of Staff, "About the Joint Chiefs of Staff," undated. As of November 2, 2016:
http://www.jcs.mil/About

Joint Chiefs of Staff, *CJCS Vision for Joint Development*, November 2005. As of November 3, 2016:
http://www.dtic.mil/doctrine/education/officer_JPME/cjcsvision_jod.pdf

Joint Chiefs of Staff, *Joint Publication 1: Doctrine for the Armed Forces of the United States*, JP-1, March 25, 2013.

Joint Forces Staff College, *The Joint Staff Officer's Guide 2000*, JFSC Pub 1, 2000. As of November 3, 2016:
http://www.au.af.mil/au/awc/awcgate/pub1/index2000.htm

Kakesako, Gregg K., "General Pulls Plug on Camp Smith Job," *Honolulu Star Bulletin*, October 7, 2004. As of January 2, 2016:
http://archives.starbulletin.com/2004/10/07/news/story1.html

Kamarck, Kristy N., *Goldwater-Nichols and the Evolution of Office Joint Professional Military Education*, Washington, D.C.: Congressional Research Service, January 13, 2016. As of November 3, 2016:
https://www.fas.org/sgp/crs/natsec/R44340.pdf

Kohn, Richard, "The Erosion of Civil Control in the United States Military Today," *Naval War College Review*, Vol. LV, No. 3, Summer 2002, pp. 8–59. As of January 4, 2016:
https://usnwc.edu/getattachment/c280d26a-9d66-466a-809b-e0804cbc05f4/
erosion-of-civilian-control-of-the-military-in-the.aspx

Kuvaas, Jay, ed., *Napoleon on the Art of War*, New York: Free Press, 1999.

Lambeth, Benjamin S., *The Transformation of American Airpower*, Ithaca, N.Y.: Cornell University Press, 2000.

Lee, Caitlin, "The Culture of US Air Force Innovation: A Historical Case Study of the MQ-1 Predator Program," doctoral dissertation, King's College, London, May 2016.

Levins, Harry, "Pace Appointment Shows Marines Have Arrived," *St. Louis Post Dispatch*, April 23, 2005, p. 23.

Lim, Nelson, Jefferson P. Marquis, Kimberly Curry Hall, David Schulker, and Xiaohui Zhuo, *Officer Classification and the Future of Diversity Among Senior Military Leaders: A Case Study of the Army ROTC*, Santa Monica, Calif.: RAND Corporation, TR-731-OSD, 2009. As of June 29, 2017:
https://www.rand.org/pubs/technical_reports/TR731.html

Locher, James R., III, "Has It Worked? The Goldwater-Nichols Reorganization Act," *Naval War College Review*, Vol. 56, No. 4, Autumn 2001, pp. 95–115. As of November 3, 2016:
https://www.usnwc.edu/getattachment/744b0f7d-4a3f-4473-8a27-c5b444c2ea27/Has-It-Worked--The-Goldwater-Nichols-Reorganization

Locher, James R., III, *Victory on the Potomac*, College Station, Tex.: Texas A&M University Press, 2004.

Luck, Gary, Gen (Ret.), Col Mike Findlay, and the Joint Warfighting Center Joint Training Division, *Joint Operations: Insights and Best Practices*, 2nd ed., July 2008. As of January 2, 2016:
http://www.au.af.mil/au/awc/awcgate/jfcom/joint_ops_insights_july_2008.pdf

Mack, Russell, Col (USAF), *Creating Joint Leaders Today for a Successful Air Force Tomorrow*, Maxwell Air Force Base, Ala.: Air University, May 11, 2010. As of November 3, 2016:
http://projects.iq.harvard.edu/files/fellows/files/mack.pdf?m=1461781498

Marks, Edward, "Rethinking the Geographic Combatant Commands," *InterAgency Journal*, Vol. 1, No. 1, Fort Leavenworth, Kan.: Colonel Arthur D. Simons Center for the Study of Interagency Cooperation, Fall 2010, pp. 19–23. As of December 21, 2016:
http://thesimonscenter.org/wp-content/uploads/2010/11/IAJ-1-1-pg19-23.pdf

Marx, Aaron, LtCol. (USMC), *Rethinking Marine Corps Officer Promotion and Retention*, Washington, D.C.: Brookings Institution, August 2014. As of November 3, 2016:
https://www.brookings.edu/wp-content/uploads/2016/06/Rethinking-Marine-Corps-Officer-Promotion-73014x2.pdf

McInnis, Kathleen, *Goldwater-Nichols at 30: Defense Reform and Issues for Congress*, Congressional Research Service, R44474, June 2, 2016. As of November 3, 2016:
https://www.fas.org/sgp/crs/natsec/R44474.pdf

Mehta, Aaron, "SASC NDAA Swerves Hard on Goldwater Nichols Reforms," *Defense News*, May 12, 2016. As of November 3, 2016: http://www.defensenews.com/story/defense/policy-budget/budget/2016/05/12/sasc-ndaa-swerves-hard-goldwater-nichols-reforms-mccain/84311626/

Meilinger, Phillip S., Col (USAF, Ret.) "Airpower Past, Present and Future," presented at the Air Force Historical Foundation Symposium, Arlington, Va., October 16–17, 2007.

Mendel, William, COL (U.S. Army, Ret.), and Graham H. Turbiville Jr., *The CINC's Strategies: The Combatant Command Process*, Fort Leavenworth, Kan.: Foreign Military Studies Office, Combined Arms Command, December 1, 1997. As of December 21, 2016: http://www.strategicstudiesinstitute.army.mil/pdffiles/PUB316.pdf

Mullen, Michael, ADM (U.S. Navy), "Joint Chiefs of Staff Speech to the Naval War College," Newport, R.I.: Naval War College, January 8, 2010.

Murdock, Clark A., Michele A. Fluornoy, Christopher A. Williams, and Kurt M. Campbell, *Beyond Goldwater-Nichols: Defense Reform for a New Strategic Era, Phase 1 Report*, Washington, D.C.: Center for Strategic and International Studies, March 2004. As of December 21, 2016: https://csis-prod.s3.amazonaws.com/s3fs-public/legacy_files/files/media/csis/pubs/bgn_ph1_report.pdf

Murray, Shoon, "Ambassadors and the Geographic Commands," paper presented at Maritime Power and International Security EMC Chair Symposium, Newport, R.I.: U.S. Naval War College, March 25–26, 2015. As of December 24, 2016: https://www.usnwc.edu/Academics/Faculty/Derek-Reveron/Workshops/Maritime-Security,-Seapower,---Trade-(1)/papers/murray.aspx

"Oral History: Barnard Trainor," *PBS Frontline*, undated As of April 10, 2017: http://www.pbs.org/wgbh/pages/frontline/gulf/oral/trainor/1.html

Owens, Mackubin Thomas, "What Military Officers Need to Know About Civil-Military Relations," *Naval War College Review*, Vol. 65, No. 2, Spring 2012. As of December 29, 2016: https://www.usnwc.edu/getattachment/1ef74daf-ebff-4aa4-866e-e1dd201d780e/What-Military-Officers-Need-to-Know-about-Civil-Mi

Palmer, Ian C., MAJ (U.S. Army), "A Model for Joint Leadership Doctrine," */luce.nt/*, 2012, pp. 40–54. As of November 3, 2016: https://www.usnwc.edu/Lucent/OpenPdf.aspx?id=143&Title=2011-2012

Pike, John, "Joint Special Operations Command (JSOC)," GlobalSecurity.org, undated. As of January 31, 2017:
http://www.globalsecurity.org/military/agency/dod/jsoc.htm

Powell, Colin, GEN (U.S. Army), "Testimony: Department of Defense Appropriations for Fiscal Year 1992," hearings before a subcommittee of the Senate Appropriations Committee, 102nd Congress, March 4, 1991.

Priest, Dana, "A Four-Star Foreign Policy?" *Washington Post*, September 28, 2000. As of November 3, 2016:
https://www.washingtonpost.com/archive/politics/2000/09/28/a-four-star-foreign-policy/f9779938-7a88-449f-9f55-84ab020abbd7/

Priest, Dana, *The Mission: Waging War and Keeping Peace with America's Military*, New York: W. W. Norton & Company, 2003.

Rearden, Steven L., *The Role and Influence of the Chairman: A Short History*, Joint History Office, Joint Chiefs of Staff, September 28, 2011. As of November 3, 2016:
http://www.jcs.mil/Portals/36/Documents/History/Institutional/The_Role_and_Influence_of_the_Chairman.pdf

Reveron, Derek S., "Shaping and Military Diplomacy," presentation at the 2007 Annual Meeting of the American Political Science Association, August 30–September 2, 2007. As of December 24, 2016:
http://www.faoa.org/resources/documents/apsa07_proceeding_210193.pdf

Ricks, Thomas E., "Army Historian Cites Lack of Postwar Plan," *Washington Post*, December 25, 2004. As of December 29, 2016:
http://www.washingtonpost.com/wp-dyn/articles/A24891-2004Dec24.html

Robbert, Al, "Air Force Competitiveness for Senior Joint Assignments: Recommendations for SECAF and CSAF," briefing, March 2009.

Rogers, Rep. Mike, "Remarks to 2017 Space Symposium," April 4, 2017. As of April 25, 2017:
http://cdn.defensedaily.com/wp-content/uploads/post_attachment/163165.pdf

Rolan, Troy A., Sr., "Department of the Army Announces Deployment of Fort Bragg–Based Units," U.S. Army, March 18, 2016. As of December 28, 2016:
https://www.army.mil/article/164519/department_of_the_army_announces_deployment_of_fort_bragg_based_units

Rumbaugh, R. Russell, "The Best Man for the Job? Combatant Commanders and the Politics of Jointness," *Joint Force Quarterly*, Vol. 75, September 30, 2014. As of November 3, 2016:
http://ndupress.ndu.edu/Media/News/News-Article-View/Article/577567/jfq-75-the-best-man-for-the-job-combatant-commanders-and-the-politics-of-jointn/

Saslow, Scott, "Inside the US Navy's Leadership School," *Forbes*, April 27, 2010. As of December 29, 2016:
http://www.forbes.com/2010/04/27/
navy-executive-learning-office-leadership-managing-education.html

Schwarzkopf, H. Norman, GEN (U.S. Army, Ret.), *It Doesn't Take a Hero*, excerpt, *Newsweek*, September 27, 1992. As of December 29, 2016:
http://www.newsweek.com/schwarzkopf-198508

Shaw, Ryan, and Miriam Krieger, "Don't Leave Jointness to the Services: Preserving Joint Officer Development Amid Goldwater-Nichols Reform," *War on the Rocks*, December 30, 2015. As of November 3, 2016:
http://warontherocks.com/2015/12/
dont-leave-jointness-to-the-services-preserving-joint-officer-development-amid-goldwater-nichols-reform/

Smith, Jeffrey, *Tomorrow's Air Force: Tracing the Past, Shaping the Future*, Bloomington, Ind.: Indiana University Press, 2014.

Spirtas, Michael, Thomas-Durell Young, and S. Rebecca Zimmerman, *What It Takes: Air Force Command of Joint Operations*, Santa Monica, Calif.: RAND Corporation, MG-777-AF, 2009. As of November 2, 2016:
http://www.rand.org/pubs/monographs/MG777.html

Svan, Jennifer H., "Gen. Breedlove Nominated to Head US European Command," *Stars and Stripes,* March 28, 2013. As of May 17, 2017:
https://www.stripes.com/news/
gen-breedlove-nominated-to-head-us-european-command-1.214018#.WR8tklLMxcA

Swift, Samuel A., Don A. Moore, Zacariah S. Sharek, and Francesca Gino, "Inflated Applicants: Attribution Errors in Performance Evaluation by Professionals," *PLOS One,* July 24, 2013.

Terry, James L., LTG (U.S. Army), "US Army Central: Operating in a Complex World," U.S. Army, October 1, 2015. As of December 28, 2016:
https://www.army.mil/article/156510

Thornhill, Paula, *The Crisis Within: America's Military and the Struggle Between Overseas and Guardian Paradigms*, Santa Monica, Calif.: RAND Corporation, RR-1420-AF, 2016. As of June 29, 2017:
https://www.rand.org/pubs/research_reports/RR1420.html

U.S. Department of Defense, *2001 Quadrennial Defense Review Report*, Washington, D.C., September 2001. As of November 3, 2016
http://archive.defense.gov/pubs/qdr2001.pdf

U.S. Department of Defense, "Townsend Takes Command of Operation Inherent Resolve," August 21, 2016. As of November 3, 2016: http://www.defense.gov/News/Article/Article/920930/townsend-takes-command-of-operation-inherent-resolve

U.S. Government Accountability Office, *DOD Needs to Reevaluate Its Approach for Managing Resources Devoted to Functional Combatant Commands*, Washington, D.C., GAO-14-439, June 2014. As of December 28, 2016: http://www.gao.gov/assets/670/664443.pdf

U.S. Government Accountability Office, *DOD Needs to Reassess Personnel Requirements for the Office of the Secretary of Defense, Joint Staff, and Military Service Secretariats*, Washington, D.C., GAO-15-10, January 2015. As of December 20, 2016: http://www.gao.gov/assets/670/667997.pdf

von Kuhl, Hermann, General (German Army), "Unity of Command Among the Central Powers," *Foreign Affairs*, September 1923. As of December 20, 2016: https://www.foreignaffairs.com/articles/russian-federation/1923-09-15/unity-command-among-central-powers

Watson, Cynthia, *Combatant Commands: Origins, Structure, and Engagements*, Santa Barbara, Calif.: Praeger, 2011.

Worden, Michael, Col (USAF), *Rise of the Fighter Generals: The Problem of Air Force Leadership*, Maxwell Air Force Base, Ala.: Air University Press, 1998.

Whittle, Richard, *Predator: The Secret Origins of the Drone Revolution*, New York: Harper Collins, 2014.

Wright, Stephen E., "Two Sides of the Coin: The Strategist and the Planner," in Richard Bailey and James Forsyth, eds., *Strategy: Context and Adaptation from Archidamus to Airpower*, Annapolis, Md.: Naval Institute Press, 2016.